W9-BKO-085

Solutions and Tests

for

Exploring Creation

with

Physical Science
2nd Edition

by
Dr. Jay L. Wile

Solutions and Tests for Exploring Creation With Physical Science, 2nd Edition

Published by
Apologia Educational Ministries, Inc.
1106 Meridian Plaza, Suite 220
Anderson, IN 46016
www.apologia.com

Copyright © 2007 Apologia Educational Ministries, Inc. All rights reserved.

Manufactured in the United States of America
Fourth Printing: April 2010

ISBN: 978-1-932012-78-1

Printed by Courier, Inc., Stoughton, MA

Cover photos: [agency: Dreamstime.com – glacier © Michael Klenetsky]
[agency: Istockphoto.com – cumulus clouds © David Raboin, aurora © Roman Krochuk],
sun and earth courtesy of NASA

Cover design by Kim Williams

Exploring Creation with Physical Science, 2ⁿᵈ Edition
Solutions and Tests

TABLE OF CONTENTS

Module Tests:

Solutions to the Module Tests:

Quarterly Tests:

Solutions to the Quarterly Tests:

TEACHER'S NOTES
Exploring Creation With Physical Science, 2nd Edition

Thank you for choosing *Exploring Creation With Physical Science*. I designed this course to meet the needs of the homeschooling parent. I understand that most homeschooling parents do not know physical science very well, if at all. As a result, they consider it nearly impossible to teach to their children. This course has several features that make it ideal for such a parent:

1. The course is written in a conversational style. Unlike many authors, I do not get wrapped up in the desire to write formally. As a result, the text is easy to read, and the student feels more like he or she is *learning*, not just reading.

2. The course is completely self-contained. Each module in the student text includes the text of the lesson, experiments to perform, problems to work, and questions to answer. This book contains the solutions to the study guides in the student text, tests, solutions to the tests, and some extra material (answers to some experiments, answers to the module summaries, cumulative tests, and solutions to the cumulative tests).

3. The experiments are designed for the home. They can be done with items that are readily available at either the grocery store or the hardware store.

4. Most importantly, this course is Christ-centered. In every way possible, I try to make physical science glorify God. One of the most important things that you and your student should get out of this course is a deeper appreciation for the wonder of God's creation!

Pedagogy of the Text

There are two types of exercises that the student is expected to complete: "On Your Own" problems and study guide questions.

- The "On Your Own" problems should be answered as the student reads the text. The act of working out these problems will cement in the student's mind the concepts he or she is trying to learn. The solutions to these problems are included as a part of the student text. The student should feel free to use those solutions to check his work.

- A study guide is found at the end of each module. It is designed to help the student review what has been covered over the course of the module. It should not be started until *after* the student has completed the module. That way, it will function as a review. It can also be used as a study aid for the test. The student should feel free to use the book while answering the study guide questions.

In addition to the exercises, there is also a test for each module. Those tests are in this book, but a packet of the tests is also included with this book. You can tear the tests out of the packet and give them to your student so that you need not give him this book to administer the tests. You can also purchase additional packets for additional students. You also have our permission to copy the tests out of this book if you would prefer to do that instead of purchasing additional tests for additional students. **I strongly recommend that you administer each test once the student has completed the module**

and the study guide. The student should be allowed to have only a calculator, pencil, and paper while taking the test.

There are also cumulative tests in this book. You can decide whether or not to give these tests to your student. Cumulative tests are probably a good idea if your student is planning to go to college, as he or she will need to get used to taking such tests. There are four cumulative tests along with their solutions. Each cumulative test covers four modules. You have three options as to how you can administer them. You can give each test individually so that the student has four quarterly tests. You can combine the first two quarterly tests and the second two quarterly tests to make two semester tests. You can also combine all four tests to make one end-of-the-year test. If you are giving these tests for the purpose of college preparation, I recommend that you give them as two semester tests, because that is what the student will face in college. The cumulative tests are not in the packet of tests. However, you have our permission to copy them out of this book so that you can give them to your student.

Any information that the student must memorize is centered in the text and put in boldface type. Any boldface words (centered or not) are terms with which the student must be familiar. In addition, all definitions presented in the text need to be memorized. Finally, any information required to answer the questions on the study guide must be committed to memory for the test. If the study guide tells the student that he can refer to a particular table or figure in the text, the test will allow him to do so as well. However, if the study guide does not specifically indicate that the student can reference a figure or table, the student will not be able to reference it for the test.

You will notice that every solution contains an underlined section. That is the answer. The rest is simply an explanation of how to get the answer. For questions that require a sentence or paragraph as an answer, the student need not have *exactly* what is in the solution. The basic message of his or her answer, however, has to be the same as the basic message given in the solution.

Experiments

The experiments in this course are designed to be done as the student is reading the text. I recommend that your student keep a notebook of these experiments. The details of how to perform the experiments and how to keep a laboratory notebook are discussed in the "Student Notes" section of the student text. If you go to the course website that is discussed in the "Student Notes" section of the student text, you will also find examples of how the student should record his or her experiments in the laboratory notebook.

Grading

Grading your student is an important part of this course. I recommend that you *correct* the study guide questions, but I do not recommend that you include the student's score in his or her grade. Instead, I recommend that the student's grade be composed solely of test grades and laboratory notebook grades. Here is what I suggest you do:

1. Give the student a grade for each lab that is done. This grade should not reflect the accuracy of the student's results. Rather, it should reflect how well the student followed directions, how well the student recorded his data, and how well he wrote up the lab in his lab notebook.

2. Give the student a grade for each test. In the test solutions, you will see a point value assigned to each problem. If your student answered the problem correctly, he or she should receive the number of points listed. If your student got a portion of the problem correct, he or she should receive a portion of those points. Your student's percentage grade, then, can be calculated as follows:

$$\text{Student's Grade} = \frac{\text{\# of points received}}{\text{\# of points possible}} \times 100$$

The number of possible points for each test is listed at the bottom of the solutions.

3. The student's overall grade in the course should be weighted as follows: 35% lab grade and 65% test grade. If you use the cumulative tests, make them worth twice as much as each module test. If you really feel that you must include the study guides in the student's total grade, make the labs worth 35%, the tests worth 55%, and the study guides worth 10%. A straight 90/80/70/60 scale should be used to calculate the student's letter grade. This is typical for most schools. If you have your own grading system, please feel free to use it. This grading system is only a suggestion.

Finally, I must tell you that this course is user-friendly and reasonably understandable. At the same time, however, *it is not EASY*. It is a tough course. Thus, a letter grade of "C" would represent the score of the average student who will most likely be college bound.

Question/Answer Service

For all those who use this curriculum, we offer a question/answer service. If there is anything in the modules that you do not understand - from an esoteric concept to a solution for one of the problems - just contact us via any of the methods listed on the **NEED HELP?** page of the student text. You can also contact us regarding any grading issues that you might have. This is our way of helping you and your student to get the maximum benefit from our curriculum.

SOLUTIONS TO THE MODULE #1 STUDY GUIDE

1. a. <u>Atom</u> – The smallest chemical unit of matter

b. <u>Molecule</u> – Two or more atoms linked together to make a substance with unique properties

c. <u>Concentration</u> – The quantity of a substance within a certain volume

2. <u>Carbon disulfide is made up of molecules.</u> Since it can be broken down into smaller components, it must be made of molecules. If it were made of atoms, it could not be broken down into smaller units, as atoms are the smallest chemical unit of matter. Although the molecules are composed of atoms, it is not correct to say carbon disulfide is made of atoms. Remember, a molecule has properties that are unique. Thus, no atom has the properties of carbon disulfide. Only carbon disulfide molecules have the properties of carbon disulfide, so it is made of molecules.

3. Rust is not attracted to a magnet because <u>when an atom is a part of a molecule, the molecule does not take on the characteristics of the atom. Instead, the atoms in the molecule join together in such a way as to give the molecule its own unique characteristics.</u>

4. <u>The statue will eventually turn a shade of green, just like the copper wire did in Experiment 1.1.</u> This comes from the copper atoms reacting with water and carbon dioxide in the air to make copper hydroxycarbonate.

5. <u>Scientists have NOT seen atoms.</u> The scanning tunneling electron microscope "pictures" that you see are not pictures of atoms. Instead, they are the result of computer calculations involving electricity and a theory called quantum mechanics.

6. The prefix <u>centi means 0.01</u>; the prefix <u>milli means 0.001</u>; and the prefix <u>kilo means 1,000</u>.

7. <u>Mass is measured in grams in the metric system. In the English system, it is measured in slugs.</u>

8. <u>Volume is measured in liters in the metric system. In the English system, it is measured in gallons, pints, or quarts.</u>

9. <u>Length is measured in meters in the metric system. In the English system, it is measured in feet, yards, inches, or miles.</u>

10. Since we want to convert from meters to centimeters, we need to remember that "centi" means "0.01." So 1 centimeter is the same thing as 0.01 meters. Thus:

$$1 \text{ cm} = 0.01 \text{ m}$$

Next, we write the measurement as a fraction:

$$\frac{1.3 \text{ m}}{1}$$

Now we can use our conversion relationship. Since we want to end up with cm in the end, we must multiply the measurement by a fraction that has meters on the bottom (to cancel the meter unit that is there) and cm on the top (so that cm is the unit we are left with). Remember, the numbers next to the units in the relationship above go with the units. Thus, since "m" goes on the bottom of the fraction, so does "0.01." Since "cm" goes on the top, so does "1."

$$\frac{1.3 \ \cancel{m}}{1} \times \frac{1 \ cm}{0.01 \ \cancel{m}} = 130 \ cm$$

Therefore, 1.3 m is the same as <u>130 cm</u>.

11. Since we want to convert from kg to grams, we need to remember that "kilo" means "1,000." So one kilogram is the same thing as 1,000 grams. Thus:

$$1 \ kg = 1,000 \ g$$

That's our conversion relationship. Next, we write the measurement as a fraction:

$$\frac{75 \ kg}{1}$$

Since we want to end up with grams in the end, we must multiply the measurement by a fraction that has kilograms on the bottom (to cancel the kg unit that is there) and grams on the top (so that grams is the unit we are left with):

$$\frac{75 \ \cancel{kg}}{1} \times \frac{1,000 \ g}{1 \ \cancel{kg}} = 75,000 \ g$$

The person's mass is <u>75,000 g</u>.

12. We use the same procedure we used in the previous two problems. Thus, I am going to reduce the length of the explanation.

$$\frac{0.500 \ \cancel{gal}}{1} \times \frac{3.78 \ L}{1 \ \cancel{gal}} = 1.89 \ L$$

There are <u>1.89 L</u> in 0.500 gallons.

13. $$\frac{100.0 \ \cancel{cm}}{1} \times \frac{1 \ in}{2.54 \ \cancel{cm}} = 39.4 \ in$$

There are <u>39.4 inches</u> in a meterstick. Note that I rounded the answer. The real answer was "39.370078740," but there are simply too many digits in that number. When you take chemistry, you will learn about significant figures, a concept that tells you where to round numbers off. For right now, don't worry about it. If you rounded at a different spot than I did, that's fine.

14. <u>Baking bread is not a dangerous activity because the ozone it produces is not concentrated enough to be dangerous.</u> Ozone is a poison, but at low enough concentrations, it does not adversely affect people. At higher concentrations, however, it can be toxic enough to kill you!

Sample Calculations for Experiment 1.2

Number of "finger" strings in the "cubit" string: 17.3 fingers
Length of tabletop in cubits: 2.3 cubits
Length of tabletop in fingers: 40 fingers
Width of tabletop in cubits: 2.5 cubits
Width of tabletop in fingers: 42.5 fingers

Length of the tabletop in cubits converted to fingers:

$$\frac{2.3 \text{ cubits}}{1} \times \frac{17.3 \text{ fingers}}{1 \text{ cubit}} = 39.79 \text{ fingers}$$

The converted value is very close to the measured value of 40 fingers.

Width of the tabletop in cubits converted to fingers:

$$\frac{2.5 \text{ cubits}}{1} \times \frac{17.3 \text{ fingers}}{1 \text{ cubit}} = 43.25 \text{ fingers}$$

The converted value is very close to the measured value of 42.5 fingers.

SOLUTIONS TO THE MODULE #2 STUDY GUIDE

1. a. <u>Humidity</u> - The moisture content of air

b. <u>Absolute humidity</u> – The mass of water vapor contained in a certain volume of air

c. <u>Relative humidity</u> – The ratio of the mass of water vapor in the air at a given temperature to the maximum mass of water vapor the air could hold at that temperature, expressed as a percentage.

d. <u>Greenhouse effect</u> – The process by which certain gases (principally water vapor, carbon dioxide, and methane) trap heat that radiates from earth

e. <u>Parts per million</u> – The number of molecules (or atoms) of a substance in a mixture for every 1 million molecules (or atoms) in that mixture

2. <u>The humidity is higher on the first day</u>. Since the person felt cooler on the second day (despite the same temperature), his sweat must have evaporated more quickly than on the first day. Thus, the first day was more humid.

3. <u>The child will add more water on the second day</u>. Since the humidity was lower on the second day, the water in the bowl will evaporate more quickly on that day.

4. <u>The water will not evaporate</u>. Since the relative humidity is 100%, the air cannot hold any more water vapor. As a result, no net evaporation will occur.

5. <u>Sweat cools you down because when it evaporates, it takes energy from your skin</u>. When energy leaves your skin, it gets cooler.

6. <u>Dry air is 78% nitrogen and 21% oxygen</u>.

7. <u>If the air had no carbon dioxide in it, the earth would be colder</u>. Since carbon dioxide is a greenhouse gas, the greenhouse effect would be weaker, leaving a cool earth. You could also answer this question by saying that plants would die of starvation.

8. <u>If there were no ozone in the air, ultraviolet light would kill a lot of living things</u>.

9. <u>If more oxygen were in the air, living things would not be as healthy and forest fires would increase in frequency and ferocity</u>.

10. <u>There is no reason to expect that the new planet will have the same temperature as earth. If it does not have essentially the same air, with all the same levels of all the greenhouse gases, it will not have the same temperature!</u> Mercury, for example, is much closer to the earth than Venus. Venus, however, is warmer than Mercury, because the greenhouse effect on Venus is very strong.

11. <u>Nitrogen makes up the majority of the air we exhale</u>. See Figure 2.5.

12. <u>We exhale more oxygen</u>. See Figure 2.5.

13. <u>No</u>. Figure 2.6 shows that the average global temperature has been reasonably constant for the past 80 years.

14. Remember, we know the relationship between ppm and percent. We can therefore just use the factor-label method to figure out the answer.

$$\frac{0.110 \text{ ppm}}{1} \times \frac{1\%}{10,000 \text{ ppm}} = 0.0000110\%$$

A concentration of 0.110 ppm is equal to <u>0.0000110%</u>.

15. Remember, we know the relationship between percent and ppm, so we can convert using the factor-label method.

$$\frac{0.023\%}{1} \times \frac{10,000 \text{ ppm}}{1\%} = 230 \text{ ppm}$$

The concentration of nitrogen oxides in this sample of air is <u>230 ppm</u>.

16. <u>The air is much cleaner today than 30 years ago</u>. See Figure 2.9.

17. <u>A cost/benefit analysis attempts to determine whether or not to take an action by determining the benefits of that action as well as the costs</u>. If the benefit outweighs the cost, the action should be taken. If not, the action should not be taken.

18. <u>A catalytic converter converts carbon monoxide in the car's exhaust to carbon dioxide</u>. The fact that most cars have these today is one of the main things responsible for reducing the carbon monoxide concentration in the air by more than 70% since 1975.

19. <u>A scrubber traps sulfur oxides in a smokestack and keeps them from being emitted into the air</u>. They are one of the reasons there has been more than a 70% decrease in sulfur oxides concentration in the air.

20. <u>Ground-level ozone is a pollutant because it is a poison, and it is where we can breathe it. Ozone in the ozone layer is not a pollutant because no one breathes that high up in the air, so its poisonous properties are unimportant. It is necessary in the ozone layer in order to block the sun's ultraviolet rays</u>.

SOLUTIONS TO THE MODULE #3 STUDY GUIDE

1. a. <u>Atmosphere</u> – The mass of air surrounding a planet

   b. <u>Atmospheric pressure</u> - The pressure exerted by the atmosphere on all objects within it

   c. <u>Barometer</u> - An instrument used to measure atmospheric pressure

   d. <u>Homosphere</u> – The lower layer of earth's atmosphere, which exists from ground level to roughly 80 kilometers (50 miles) above sea level

   e. <u>Heterosphere</u> – The upper layer of earth's atmosphere, which exists higher than roughly 80 kilometers (50 miles) above sea level

   f. <u>Troposphere</u> – The region of the atmosphere that extends from ground level to roughly 11 kilometers (7 miles) above sea level

   g. <u>Stratosphere</u> – The region of the atmosphere that spans altitudes of roughly 11 kilometers to 48 kilometers (30 miles)

   h. <u>Mesosphere</u> – The region of the atmosphere that spans altitudes of roughly 48 kilometers to 80 kilometers (50 miles)

   i. <u>Jet streams</u> – Narrow bands of high-speed winds that circle the earth, blowing from west to east

   j. <u>Heat</u> – Energy that is transferred as a consequence of temperature differences

   k. <u>Temperature</u> – A measure of the energy of random motion in a substance's molecules

   l. <u>Thermosphere</u> – The region of the atmosphere between altitudes of roughly 80 kilometers and 460 kilometers

   m. <u>Exosphere</u> – The region of the atmosphere above an altitude of roughly 460 kilometers

   n. <u>Ionosphere</u> – The region of the atmosphere between the altitudes of roughly 65 kilometers and 330 kilometers, where the gases are ionized

2. <u>Atmospheric pressure would be greater than it is now</u>. After all, if there were twice as many molecules in the air, the mass of air pressing down on everything in the atmosphere would be twice as high.

3. <u>The students used different liquids</u>. A given volume of the liquid used by the first student weighs more than the same volume of the liquid used by the second student. Remember how a barometer works. The height of the column of liquid is determined by the amount of liquid necessary to counteract the atmospheric pressure pushing on the liquid. The heavier the liquid, the less will be necessary to achieve this effect. Thus, if a given volume of liquid used by the first student weighs more than the same volume of liquid used by the second student, the liquid in the first student's barometer will not have to rise as high to counteract the force provided by atmospheric pressure.

4. <u>An atmospheric pressure of 25.4 inches of mercury would be reported.</u> Since 1.0 atm corresponds to the average sea-level value of atmospheric pressure, 0.85 atms means that the atmospheric pressure is lower than average.

5. <u>The first came from the homosphere.</u> In the homosphere, the mixture of gases in the air is the same throughout. It is the mixture we learned in the previous module. The heterosphere has many different compositions, depending on altitude.

6. <u>The balloon enters the stratosphere when its temperature readings cease to decrease and begin increasing. The balloon enters the mesosphere when the temperature readings cease increasing and begin decreasing again.</u> Since the temperature gradient changes at the stratosphere and then again at the mesosphere, this can be used to determine when the balloon has reached those parts of the atmosphere.

7. <u>Troposphere, stratosphere, mesosphere</u>

8. <u>I am referring to the "amount gradient."</u> You could also answer with "pressure gradient." Both the amount of air and the pressure decrease with increasing altitude. Remember, "gradient" just means steady change, so I can use that term with any quantity.

9. <u>The plane is flying in the troposphere.</u> That's where the majority of weather phenomena exist.

10. <u>The second vial contains the gas with the highest temperature.</u> Remember, temperature is a measure of the energy of random motion in a substance. Since the molecules in the second vial have a higher speed, they have more energy and thus a higher temperature.

11. <u>Your companion is correct. Heat is energy that is being transferred. The reason you are cold is that energy is being transferred from your body to the surrounding air.</u> Even though it sounds weird to say it, you get cold because of transferred energy; thus, you get cold because of heat!

12. <u>The temperature gradient would reverse, getting warmer near that region.</u> Remember, the temperature increases with increasing altitude in the stratosphere because of a layer of the greenhouse gas ozone. Carbon dioxide is also a greenhouse gas, and thus would produce roughly the same effect.

13. <u>A ban on CFCs will probably not save or improve lives because CFCs cause a depletion of ozone only during a few months out of the year and mostly over Antarctica.</u> Since there is no significant population there, and since the depletion is temporary, the "ozone hole" is not a big threat to human survival.

14. <u>A ban on CFCs will most likely cost many lives because refrigeration, surgical sterilization, and firefighting will all be less efficient, causing death by starvation, death by eating food-borne illness, death by surgical infection, and death by fire.</u>

15. <u>Some kinds of human-made molecules that contain chlorine can survive the trip up to the ozone layer, while most naturally produced chlorine-containing molecules cannot.</u> Thus, although we produce few chlorine-containing molecules, many of them can reach the ozone layer, where ozone depletion can occur. As a result, most of the ozone-destroying molecules in the ozone layer are from human sources.

16. <u>The polar vortex lifts the CFCs into the ozone layer.</u> Since the polar vortex is seasonal and limited mostly to the South Pole, so is ozone depletion.

17. <u>The ionosphere is a stretch of the atmosphere ranging from the upper mesosphere to the lower parts of the thermosphere. It is useful to us in radio communication, as radio signals can bounce off of it to extend their range.</u> An altitude range of roughly 65 km to 330 km is also a valid answer to where the ionosphere is.

SOLUTIONS TO THE MODULE #4 STUDY GUIDE

1. a. <u>Electrolysis</u> – The use of electricity to break a molecule down into smaller units

b. <u>Polar molecule</u> – A molecule that has slight positive and negative charges due to an imbalance in the way electrons are shared

c. <u>Solvent</u> – A liquid substance capable of dissolving other substances

d. <u>Solute</u> – A substance that is dissolved in a solvent

e. <u>Cohesion</u> – The phenomenon that occurs when individual molecules are so strongly attracted to each other that they tend to stay together, even when exposed to tension

f. <u>Hard water</u> - Water that has certain dissolved ions in it – predominately calcium and magnesium ions

2. <u>The result would be equal amounts of oxygen and hydrogen (option b)</u>. The chemical formula says that there are 2 hydrogens and 2 oxygens in each molecule of hydrogen peroxide. Thus, there are equal amounts of each atom, resulting in equal amounts of each gas.

3. <u>The chemical formula HO would be the more likely erroneous result</u>. If the test tube that held hydrogen had a slow leak, it would look like less hydrogen was collected than what should have been collected. Thus, the experiment would indicate a chemical formula with *less* hydrogen atoms in it. H_4O is an erroneous result that indicates there were *more* hydrogens, HO is the one that indicates *less* hydrogens.

4. There are no subscripts after the Mg or the S, indicating <u>one magnesium atom and one sulfur atom</u>. There is a subscript of 4 after the O, indicating <u>four oxygen atoms</u>.

5. You put the subscripts after each symbol to indicate the number of atoms. If the number is one, there is no subscript. This leads to an answer of <u>$CaCO_3$</u>.

6. There is no subscript after the N, indicating one atom there. The next letter is capital, so it must represent another atom. There is a subscript of 3 after it, indicating three of those. Thus, there are a total of <u>four</u> atoms.

7. <u>No, the molecule will be not be polar</u>. If the atoms all pull on electrons with the same strength, none will be able to get more than its fair share.

8. <u>Baking soda will not dissolve in vegetable oil</u>. Since baking soda dissolves in water, it is either ionic or polar. (It is ionic.) Either way, it will not dissolve in a nonpolar substance because only other nonpolar substances will dissolve in nonpolar substances.

9. <u>Carbon tetrachloride is probably made of nonpolar molecules</u>, otherwise it probably would have dissolved in water.

10. <u>The liquid would have more molecules</u>, because the molecules in solid water are farther apart than they are in liquid water. For the same volume, then, there would be more molecules in the liquid.

11. <u>For nearly any other substance, the answer would be that the solid would have more molecules.</u> For almost all other substances, molecules are closer together in the solid state, so there would be more molecules in an equal volume as compared to the liquid.

12. <u>Hydrogen bonding is responsible</u>. Hydrogen bonding brings the molecules close together and makes them want to stay close together.

13. <u>Cohesion causes surface tension</u>. You could say hydrogen bonding here if you are talking about water. However, other substances exhibit surface tension, even if they do not hydrogen bond. Hydrogen bonding just makes water's surface tension larger than that of many other substances.

14. <u>Water is harder in some regions of the world because there is a higher amount of metal-containing substances in some regions than others</u>. It is the dissolved metal ions that cause hard water.

SOLUTIONS TO THE MODULE #5 STUDY GUIDE

1. a. Hydrosphere – The sum of all water on a planet

b. Hydrologic cycle – The process by which water is continuously exchanged between earth's various water sources

c. Transpiration – Evaporation of water from plants

d. Condensation – The process by which a gas turns into a liquid

e. Precipitation – Water falling from the atmosphere as rain, snow, sleet, or hail

f. Distillation – Evaporation and condensation of a mixture to separate out the mixture's individual components

g. Residence time – The average time a given particle will stay in a given system

h. Salinity – A measure of the mass of dissolved salt in a given mass of water

i. Firn – A dense, icy pack of old snow

j. Water table – The line between the water-saturated soil and the soil that is not saturated with water

k. Percolation – The process by which water moves downward in the soil, toward the water table

l. Adiabatic cooling – The cooling of a gas that happens when the gas expands with no way of getting more energy

m. Cloud condensation nuclei – Small airborne particles upon which water vapor condenses to form clouds

2. The vast majority of water on the earth is saltwater, since more than 97% of earth's water supply is in the oceans.

3. The largest source of freshwater is the glaciers and icebergs on the planet.

4. The largest source of liquid freshwater is groundwater.

5. Water can enter the atmosphere through evaporation and transpiration.

6. If the raindrop never really soaks into the soil, it can end up in a river via surface runoff. It could also soak into the groundwater and get to the river via groundwater flow. Alternatively, it could go into the soil, be absorbed by a plant, transpired into the atmosphere, condensed into a cloud, and precipitated into the river. It could also be evaporated before it soaks into the ground, condensed into a cloud, and precipitated into the river. That's four answers, but you only need three of them.

7. <u>Transpiration</u> takes water from the soil and puts it in the atmosphere, because plants absorb the soil moisture and then put it into the atmosphere via transpiration.

8. <u>The residence time in the river is shorter</u>. The residence time will be shorter wherever water is quickly exchanged with other sources.

9. <u>A lake must have a way to get rid of water other than just evaporation</u>. This usually is accomplished when the lake feeds a river or stream. If evaporation is the only way of getting rid of water, the salts that the lake receives will become concentrated, making saltwater.

10. <u>The oceans are not salty enough for the earth to be billions of years old</u>. Since salt accumulates in the oceans, the older the earth is, the saltier the oceans will be. Calculations indicate that even assuming the oceans had no salt to begin with, it would take, at most, 62 million years (*not billions of years!*) to make the oceans as salty as they are now.

11. <u>Melted sea ice would taste like saltwater</u>, because salt is incorporated into sea ice when it freezes.

12. <u>Icebergs come from glaciers</u>. If a glacier moves to the sea, it can break apart and float away as icebergs.

13. <u>Glaciers start on mountains as the result of snow that never completely melts in the summer</u>. If enough snow piles up, the weight causes it to slide down the mountain as a glacier.

14. When a piece of a glacier breaks up and falls into the ocean, we called it <u>calving</u>.

15. <u>The captain is worried because 90% of the glacier is underwater and therefore not visible.</u> The captain steered clear of the visible part, but the underside of the boat could still hit the part that is underwater.

16. <u>The water table will be deeper in the area with lots of trees</u>. Since there are no trees to take away soil moisture in the one area, and since they each have the same kind of grass, the area with the trees depletes soil moisture faster than the other one. As a result, there will be more unsaturated soil in the region with trees, and the water table will therefore be deeper.

17. <u>The air will cool as it expands</u>. That's what adiabatic cooling is all about.

18. Like the cloud in Experiment 5.3, <u>the fog will be thicker in the smoky area</u>.

19. Like the cloud in Experiment 5.3, <u>adiabatic</u> cooling accounts for most cloud formation.

20. A refrigerator compresses a gas, which heats the gas up, and then it allows the gas to expand and any liquid to evaporate, which cools the gas. The only reason the inside of the refrigerator is cold is that the pipes carrying the expanded gas are exposed to the inside of the refrigerator. <u>If you simply reversed the design so that the pipes carrying the gas after compression are on the inside of the refrigerator, the inside would heat up</u>.

21. <u>Groundwater pollution</u> is hard to track back because there is no easy way to tell where the polluted groundwater came from.

SOLUTIONS TO THE MODULE #6 STUDY GUIDE

1. a. <u>Earth's crust</u> – Earth's outermost layer of rock

b. <u>Sediment</u> – Small, solid fragments of rock and other materials that are carried and deposited by wind, water, or ice. Examples would be sand, mud, or gravel.

c. <u>Sedimentary rock</u> – Rock formed when chemical reactions cement sediments together, hardening them

d. <u>Igneous rock</u> – Rock that forms from molten rock

e. <u>Metamorphic rock</u> – Igneous or sedimentary rock that has been changed into a new kind of rock as a result of great pressure and temperature

f. <u>Plastic rock</u> – Rock that behaves like something between a liquid and a solid

g. <u>Earthquake</u> – Vibration of the earth that results either from volcanic activity or rock masses suddenly moving along a fault

h. <u>Fault</u> – The boundary between two sections of rock that can move relative to one another

i. <u>Focus</u> – The point where an earthquake begins

j. <u>Epicenter</u> – The point on the surface of the earth directly above an earthquake's focus

2. The earth is divided into the <u>atmosphere, hydrosphere, crust, mantle, and core.</u>

3. We can directly observe the <u>atmosphere, hydrosphere, and crust.</u>

4. <u>The Moho separates the crust from the mantle, and the Gutenberg discontinuity separates the mantle from the core. The Lehmann discontinuity separates the inner core from the outer core.</u>

5. <u>Sedimentary rock is formed when sediments are solidified through chemical reactions. Igneous rock forms when molten rock solidifies.</u>

6. <u>Metamorphic rock</u> starts out as either igneous or sedimentary rock and is then transformed by high temperature and pressure.

7. The extremes in temperature and pressure make the <u>rock behave sometimes like a liquid and sometimes like a solid.</u> That's why we call it plastic rock.

8. <u>Scientists observe seismic waves</u>, which are usually generated by earthquakes. The behavior of these waves tells us a lot about the makeup and properties of the mantle and core.

9. <u>The inner core is solid because of pressure freezing.</u> Even though the inner core is hotter than the outer core, it remains solid because the pressure is so great that it forces iron atoms close enough together to be solid.

10. The magnetic field is generated in the earth's core.

11. The magnetic field is caused by a large amount of electrical flow in the core.

Start 12. The dynamo theory says that the motion of the core is due to temperature differences in the core and the rotation of the earth. This motion causes the motion of electrical charges in the core, which creates electrical current. The rapid-decay theory states that the electrical current in the core started as a consequence of how the earth formed and is decreasing over time.

13. The rapid-decay theory has been used to accurately predict the magnetic fields of other planets. The dynamo theory fails miserably at this.

14. The rapid-decay theory requires a global catastrophe in order to be consistent with the data that indicate the magnetic field of the earth has reversed several times.

15. The fact that the rapid-decay theory requires a catastrophe like the worlwide Flood and the fact that the rapid-decay theory indicates an earth 10,000 years old or younger tend to make many scientists shy away from it. This is unfortunate, as there are good reasons to believe both of them!

16. Without the magnetic field, cosmic rays from the sun would hit the earth. These rays would kill all life on the planet.

17. The plates are large "islands" of the earth's lithosphere. These plates float around on the plastic rock of the asthenosphere.

18. One plate can slide under another and form a trench; the plates can move away from each other, allowing magma to rise and create new crust; the plates can push against each other, causing the crust to fold; or the plates can slide along each other.

19. Pangaea is a hypothetical supercontinent that might have existed in earth's past. At one time, all the continents might have fit together to form this supercontinent.

20. Some good scientists ignore plate tectonics because it is typically linked to the idea of an earth that is billions of years old. This is unfortunate because there is no reason to believe that the continents always moved slowly. Indeed, in a catastrophe like a worldwide Flood, they could have moved very quickly.

21. Earthquakes are caused by the motion of rock masses along a fault or by volcanic activity.

22. In the elastic rebound theory, rock masses moving relative to one another get caught on the rough, jagged edges of the fault that lies between them. The rock masses start to bend as they keep trying to move. At some point, the stress becomes too great, and the moving rock breaks free, causing the rock masses on both sides of the fault to snap back into their original shapes. The resulting vibrations are what we feel as an earthquake.

23. For every one step up in the Richter scale, the energy of the earthquake multiplies by 32. The first earthquake measured 4, and the second measured 8. The second earthquake was 4 units higher, which means it released 32 x 32 x 32 x 32 = 1,048,576 times more energy than the first!

24. <u>The four types of mountains are: volcanic mountains, domed mountains, fault-block mountains, and folded mountains. Volcanic and domed mountains need magma from the earth's mantle, fault-block mountains need vertical motion along a fault, and folded mountains need rock masses pushing against each other</u>.

SOLUTIONS TO THE MODULE #7 STUDY GUIDE

1. a. <u>Aphelion</u> – The point at which the earth is farthest from the sun

b. <u>Perihelion</u> – The point at which the earth is closest to the sun

c. <u>Lines of longitude</u> – Imaginary lines that run north and south across the earth

d. <u>Lines of latitude</u> – Imaginary lines that run east and west across the earth

e. <u>Coriolis effect</u> – The way in which the rotation of the earth bends the path of winds, sea currents, and objects that fly through different latitudes

f. <u>Air mass</u> – A large body of air with relatively uniform pressure, temperature, and humidity

g. <u>Weather front</u> – A boundary between two air masses

2. <u>The weather changes from day to day, while the climate does not</u>. Climate is what you generally expect from a region, while weather is what actually happens from day to day.

3. The three main factors are <u>thermal energy, uneven distribution of thermal energy, and water vapor in the atmosphere</u>.

4. There is no answer for this one. Just be sure that given a picture or drawing of a cloud, you can determine which of the four major types of cloud it is.

5. When a cloud is dark, you add a suffix of "nimbus" or prefix of "nimbo." The proper term is <u>nimbostratus</u>, but stratonimbus is also correct.

6. When a cloud is higher than usual, you add the "alto" prefix. Thus, it would be an <u>altolenticular cloud</u>.

7. Insolation stands for <u>incoming solar radiation</u>.

8. In the Northern Hemisphere, June 21 is the summer solstice. That's when the days in the Northern Hemisphere are at their longest, because the Northern Hemisphere is pointed towards the sun. They then begin to decrease so that by September 22 (the autumnal equinox), they are exactly 12 hours long. Thus, <u>the days are greater than 12 hours long but are decreasing in length</u>. Note that June 22 and September 23 can also be used.

9. In the Southern Hemisphere, the days get longer from June 21 (the Northern Hemisphere's summer solstice) to the December 21 (the Northern Hemisphere's winter solstice). On September 22 (the Northern Hemisphere's autumnal equinox), they are exactly 12 hours long. Thus, <u>from June 21 to September 22</u>, the days get longer but are still under 12 hours. Note that June 22 and September 23 can also be used.

10. The Northern Hemisphere is pointed toward the sun at <u>aphelion</u>, so that's when it's summer in that hemisphere.

11. <u>Temperature differences cause winds</u>.

12. There isn't a steady stream of wind blowing from the poles to the equator because <u>the temperature of the air changes as it changes latitude</u>. This causes loops of wind to develop at different latitudes.

13. <u>The Coriolis effect bends the wind patterns</u>.

14. The ground in Alaska rotates slowly compared to the ground at the equator. Thus, as the missile travels, the ground underneath it will outrun it. As a result, the missile's path will bend to the west relative to the ground. To correct for this, you will have to fire the missile <u>southeast</u>.

15. Along the surface of the earth, winds blow from cold to warm. Thus, <u>the wind will blow from the mountain to the valley</u>.

16. Since it is continental, the <u>humidity is low</u>. Since it is polar, the <u>air mass is cold</u>.

17. Since it is maritime, the <u>humidity is high</u>. Since it is tropical, the <u>air mass is warm</u>.

18. This kind of weather is indicative of a <u>warm front</u>. It is not a stationary front because the rain would have lasted several days.

19. This kind of cloud progression is caused by an <u>occluded front</u>.

20. This kind of cloud pattern and resulting rain is indicative of a cold front. Thus, <u>you should expect cooler temperatures</u>.

SOLUTIONS TO THE MODULE #8 STUDY GUIDE

1. a. <u>Updraft</u> – A current of rising air

b. <u>Insulator</u> – A substance that does not conduct electricity very well

2. <u>The Bergeron process begins with cold clouds, while the collision-coalescence theory begins with warm clouds.</u>

3. <u>The Bergeron process describes precipitation from the top of cumulonimbus clouds, while the collision-coalescence theory describes precipitation from nimbostratus.</u> Remember, the top of a cumulonimbus cloud is near the top of the troposphere, where water freezes. Nimbostratus clouds, however, are much lower, so they are the most likely to be warm.

4. <u>Only the size of the raindrop.</u> Drizzle has very small water droplets, while raindrops are bigger.

5. <u>Sleet is much smaller than hail, but both of them are frozen before they hit the ground. Freezing rain, on the other hand, is liquid until it hits a cold surface.</u> Hail and sleet also form differently, since hail is recycled through the cloud several times while sleet is not.

6. <u>The dew point is coldest on the second morning.</u> It takes a colder temperature to form dew from air that is less humid or is lower in pressure.

7. <u>The first stage is the cumulus stage, where there is only an updraft and no precipitation. In the second stage, the mature stage, there are updrafts, downdrafts, and precipitation. The last stage, the dissipation stage, has only downdrafts and precipitation.</u>

8. <u>The thunderstorm is probably made up of several cells.</u> The mature stage of a typical thunderstorm cell lasts no longer than 30 minutes.

9. <u>The charge imbalance first forms in the cumulonimbus cloud, and it is due to water droplets or ice crystals rubbing against each other in glancing collisions.</u>

10. <u>The return stroke is responsible for the majority of light and sound in a lightning bolt.</u>

11. <u>Thunder is the result of superheated air traveling out from the lightning bolt in waves.</u> When those waves hit our eardrum, we interpret them as sound. Since the waves are violent, the sound is loud.

12. <u>Lighting strikes tall things because the positive charges in the ground tend to pile up in a tall object,</u> since that's how they can get closest to the cloud.

13. <u>Sheet lightning is cloud-to-cloud lightning while lightning bolts are cloud-to-ground lightning.</u> The lightning bolts, therefore, hit the ground, while sheet lightning never does.

14. <u>A cumulonimbus cloud must be present to form a tornado.</u> The vortex will not form without the strong updraft of a thunderstorm cell that forms a cumulonimbus cloud.

15. The stages of a tornado are: the whirl stage, the organizing stage, the mature stage, the shrinking stage, and the decaying stage. The tornado is most destructive in its mature stage.

16. A hurricane starts out as a tropical disturbance, then becomes a tropical depression, then a tropical storm, and finally a tropical cyclone. The wind speeds in the storm determine in which classification a storm belongs.

17. Within the eye, a hurricane is calm. It is often sunny as well.

18. The Coriolis effect causes hurricanes in different hemispheres to rotate differently.

19. The atmospheric pressures are nearly equivalent. Even though they are far away from each other, they are very close to the same isobar, indicating nearly equal pressure.

20. The atmospheric pressure in Houston is lower. Houston is 3 isobars from the "L" symbol, while Atlanta is 4 isobars away. This means Atlanta's pressure is higher.

21. The occluded front has triangles and ovals on the same side. That's nearest San Francisco.

22. The warm front has only ovals on it, and the ovals point in the direction of travel. Thus, Indianapolis will get warmer weather soon.

23. Houston, TX is near a cold front, so it might have thunderstorms right now.

24. San Francisco, CA is behind an occluded front. Since the weather described is that of an occluded front, San Francisco might have just experienced such weather.

SOLUTIONS TO THE MODULE #9 STUDY GUIDE

1. a. <u>Reference point</u> – A point against which position is measured

b. <u>Vector quantity</u> – A physical measurement that contains directional information

c. <u>Scalar quantity</u> – A physical measurement that does not contain directional information

d. <u>Acceleration</u> – The time rate of change of an object's velocity

e. <u>Free fall</u> – The motion of an object when it is falling solely under the influence of gravity

2. In order for motion to occur, an object's position must change. Since this object's position is not changing, <u>it is not moving relative to the reference point.</u>

3. <u>The glass of water is moving relative to many reference points.</u> To someone standing still in the house, the glass is not moving. However, relative to any object not on earth, it is in motion. In fact, if you walk towards the glass, the glass is in motion relative to you, because the glass's position relative to you changes!

4. a. <u>The child is in motion relative to the two girls.</u> Even though the child is floating motionless, his position relative to the girls is changing. Thus, he is in motion relative to the girls.

b. <u>The first girl is in motion relative to the child.</u> Since the position of her relative to the child is changing, she is in motion relative to him.

c. <u>The second girl is motionless relative to the first girl.</u> The girls are keeping perfect pace. Thus, their positions relative to each other does not change. They are therefore both motionless with respect to each other.

5. This problem gives us distance and time and asks for speed. Thus, we need to use Equation (9.1). The problem wants the answer in miles per hour, however. We are given the time in minutes. Thus, we must make a conversion first:

$$\frac{30 \text{ minutes}}{1} \times \frac{1 \text{ hour}}{60 \text{ minutes}} = 0.5 \text{ hours}$$

Now we can use our speed equation:

$$\text{speed} = \frac{10 \text{ miles}}{0.5 \text{ hours}} = 20 \frac{\text{miles}}{\text{hour}}$$

6. This is another speed problem, but in this case, both of the units are wrong. We need meters per second, but we have kilometers and minutes. Thus, we need to make two conversions:

$$\frac{6 \text{ kilometers}}{1} \times \frac{1,000 \text{ meters}}{1 \text{ kilometer}} = 6,000 \text{ meters}$$

$$\frac{45 \text{ minutes}}{1} \times \frac{60 \text{ seconds}}{1 \text{ minute}} = 2,700 \text{ seconds}$$

Now we can use *those* numbers in our speed equation:

$$\text{speed} = \frac{6,000 \text{ meters}}{2,700 \text{ seconds}} = 2.2 \, \underline{\frac{\text{meters}}{\text{second}}}$$

7. a. This is a <u>scalar</u> quantity since it has no direction. Meters is a <u>distance</u> unit.

b. This is a <u>vector</u> quantity, because it has direction in it. The units are distance over time squared, which is <u>acceleration</u>.

c. This is a <u>scalar</u> quantity. It has no direction. The units indicate it is <u>speed</u>.

d. This is a <u>scalar</u> quantity. It has no direction. Liters is a volume unit, so it is <u>none of these</u>.

e. This is a <u>vector</u> quantity, because it has direction in it. The units are distance over time, which is <u>velocity</u>. It is not speed because speed is not a vector quantity.

f. This is a <u>scalar</u> quantity. It has no direction. The units indicate it is <u>speed</u>.

8. As the picture shows, the car is behind the truck, but they are both traveling in the same direction. Thus, we get their relative velocity by subtracting their individual speeds:

relative velocity = 57 miles per hour - 45 miles per hour = 12 miles per hour

Since the car is traveling faster than the truck, it is catching up to the truck. Thus, the relative velocity is <u>12 miles per hour towards each other</u>.

9. Since the velocity is not changing, <u>the acceleration is zero</u>. The time was just put in there to fool you. Remember, acceleration is the change in velocity. With no change in velocity, there is no acceleration.

10. The initial velocity is 0, and the final velocity is 12 meters per second east. The time is 2 seconds. This is a straightforward application of Equation (9.2).

$$\text{acceleration} = \frac{\text{final velocity} - \text{initial velocity}}{\text{time}}$$

$$\text{acceleration} = \frac{12 \, \frac{\text{meters}}{\text{second}} - 0 \, \frac{\text{meters}}{\text{second}}}{2 \text{ seconds}} = \frac{12 \, \frac{\text{meters}}{\text{second}}}{2 \text{ seconds}} = 6 \, \frac{\text{meters}}{\text{second}^2}$$

Since the speed increased, the acceleration and velocity are in the same direction. Thus, the acceleration is <u>6 m/sec^2 east</u>.

11. This is another application of Equation (9.2), because we are given time (12 minutes), initial velocity (30 miles per hour south) and final velocity (0, because it stops). We can't use the equation yet, however, because our time units do not agree. We'll fix that first:

$$\frac{12 \text{ minutes}}{1} \times \frac{1 \text{ hour}}{60 \text{ minutes}} = 0.2 \text{ hours}$$

Now that we have all time units in agreement, we can really use the acceleration equation:

$$\text{acceleration} = \frac{\text{final velocity} - \text{initial velocity}}{\text{time}}$$

$$\text{acceleration} = \frac{0 \frac{\text{miles}}{\text{hour}} - 30 \frac{\text{miles}}{\text{hour}}}{0.2 \text{ hours}} = \frac{-30 \frac{\text{miles}}{\text{hour}}}{0.2 \text{ hours}} = -150 \frac{\text{miles}}{\text{hour}^2}$$

The negative sign means the acceleration is in the opposite direction as the velocity. Thus, the acceleration is 150 miles/hour^2 north.

12. The physicist is technically correct because for an object to be in free fall, it can only be influenced by gravity. Air resistance is a second influence, and all objects experience air resistance.

13. Even though the physicist is technically correct, the effect of air resistance is so small on heavy objects that it usually can be ignored.

14. Neither will hit first. They both hit together. Remember, gravity accelerates all objects the same. Without air, there is no air resistance, so both objects are in true free fall. As a result, they will fall at exactly the same speed.

15. The rock is in free fall, so we can use Equation (9.3). Since the problem wants the answer in meters, we need to use 9.8 meters per second^2 as the acceleration.

$$\text{distance} = \frac{1}{2} \times (\text{acceleration}) \times (\text{time})^2$$

$$\text{distance} = \frac{1}{2} \times (9.8 \frac{\text{meters}}{\text{second}^2}) \times (4.1 \text{ seconds})^2 = \frac{1}{2} \times (9.8 \frac{\text{meters}}{\text{second}^2}) \times (4.1 \text{ seconds}) \times (4.1 \text{ seconds})$$

$$\text{distance} = \frac{1}{2} \times (9.8 \frac{\text{meters}}{\text{second}^2}) \times (16.81 \text{ second}^2) = 82.4 \text{ meters}$$

16. The rock is in free fall, so we can use Equation (9.3). Since the problem wants the answer in feet, we need to use 32 feet per second^2 as the acceleration.

$$\text{distance} = \frac{1}{2} \times (\text{acceleration}) \times (\text{time})^2$$

$$\text{distance} = \frac{1}{2} \times (32 \; \frac{\text{feet}}{\text{second}^2}) \times (7 \; \text{seconds})^2 = \frac{1}{2} \times (32 \; \frac{\text{feet}}{\text{second}^2}) \times (7 \; \text{seconds}) \times (7 \; \text{seconds})$$

$$\text{distance} = \frac{1}{2} \times (32 \; \frac{\text{feet}}{\cancel{\text{second}^2}}) \times (49 \; \cancel{\text{second}^2}) = \underline{784 \; \text{feet}}$$

17. Since the object covers less ground in each time interval, it is traveling slower in each time interval. This clearly means that the object is slowing down. When an object is slowing, its acceleration is in the <u>opposite direction</u> of its velocity.

Sample Calculations for Experiment 9.3

Time for the ball to fall: 0.79 seconds, 0.60 seconds, 0.70 seconds, 0.82 seconds, 0.65 seconds, 0.89 seconds, 0.83 seconds, 0.69 seconds, 0.59 seconds, 0.92 seconds

Average time for the ball to fall: (0.79 seconds + 0.60 seconds + 0.70 seconds + 0.82 seconds + 0.65 seconds + 0.89 seconds + 0.83 seconds + 0.69 seconds + 0.59 seconds + 0.92 seconds) ÷ 10 = 0.748 seconds

Distance the ball fell:

$$\text{distance} = \frac{1}{2} \times (\text{acceleration}) \times (\text{time})^2$$

$$\text{distance} = \frac{1}{2} \times (32 \; \frac{\text{feet}}{\text{second}^2}) \times (0.748 \; \text{seconds})^2 = \frac{1}{2} \times (32 \; \frac{\text{feet}}{\text{second}^2}) \times (0.748 \; \text{seconds}) \times (0.748 \; \text{seconds})$$

$$\text{distance} = \frac{1}{2} \times (32 \; \frac{\text{feet}}{\cancel{\text{second}^2}}) \times (0.559504 \; \cancel{\text{second}^2}) = 8.95 \; \text{feet}$$

Measured height of the ceiling: 9.33 feet

The height determined by the falling ball is similar to that measured with the tape measure.

SOLUTIONS TO THE MODULE #10 STUDY GUIDE

1. a. <u>Inertia</u> – The tendency of an object to resist changes in its velocity

b. <u>Friction</u> – A force that opposes motion, resulting from the contact of two surfaces

c. <u>Kinetic friction</u> – Friction that opposes motion once the motion has already started

d. <u>Static friction</u> – Friction that opposes the initiation of motion

2. <u>Newton's First Law</u> – An object in motion (or at rest) will tend to stay in motion (or at rest) until it is acted upon by an outside force.

<u>Newton's Second Law</u> – When an object is acted on by one or more outside forces, the total force is equal to the mass of the object times the resulting acceleration

<u>Newton's Third Law</u> – For every action, there is an equal and opposite reaction.

3. Newton's First Law of Motion tells us that an object will not change velocity until acted on by an outside force. Often, this force is friction. In this problem, once the ball is thrown, no forces (not even friction) are operating on the ball. Thus, even in a year, its velocity will still be <u>3.0 meters per second to the west.</u>

4. <u>The beanbag will not fall next to the tree. Instead, it will fall north of the tree</u>. This is once again an application of Newton's First Law. While it is in the boy's hand, the beanbag has a velocity going north. When the boy drops the beanbag, it will still have a velocity going north. Thus, as it falls, it will travel north. When it lands, then, it will be north of the tree. In fact, ignoring air resistance, when it hits the ground, it will be right next to wherever the boy is at that instant, because it will be traveling north with the boy's velocity.

5. <u>The beanbag will land next to the tree</u>. In this case, the beanbag has no initial velocity. It is at rest with the boy standing next to the tree. When the running boy taps the beanbag lightly, it simply falls to the ground.

6. <u>The boxes will slam into the front seat</u>. The boxes have the same velocity as the car. When the car stops, they continue to move with the same velocity. This makes them move forward relative to the car, slamming them into the front seat.

7. Remember, friction is caused by molecules on each surface attracting one another, and the strength of the attraction depends on how close they can get to each other. When the road gets wet, the grooves in the road get filled with water. This makes it harder for the bumps on the tires to fit into them, which makes it hard for the molecules to get close to one another. Thus, <u>the water fills in the grooves in the road, reducing how close the tire molecules can get to the road molecules</u>. This can become an even bigger problem when a film of water gets trapped under the tires, causing the tires to lose contact with the road. Essentially, they are traveling on the water, not the road. This situation is called "hydroplaning," and it causes the tire molecules to be so far from the road molecules that very little friction exists.

8. <u>The static frictional force is greater than the kinetic frictional force.</u> When the refrigerator is not moving, the man must overcome static friction to get it moving. Once it is moving, the man only needs to overcome the kinetic frictional force.

9. Since the object is moving with a constant velocity, we know its acceleration is *zero*. Since the total force exerted on an object is equal to the object's mass times its acceleration (Newton's Second Law), then the total force on the object is zero as well. This means that the child exerts enough force to counteract kinetic friction, but no more. We must be talking about kinetic friction, because the toy is already moving. Thus, the child exerts a force of <u>10 Newtons to the east</u>.

10. Since we know the child's mass and acceleration, we can calculate the total force acting on the child.

$$\text{total force} = (\text{mass}) \cdot (\text{acceleration})$$

$$\text{total force} = (20 \text{ kg}) \cdot \left(2.0 \frac{\text{m}}{\text{sec}^2} \right) = 40 \text{ Newtons}$$

Since we are ignoring friction, the only force involved is the force that the father exerts. Thus, the total force is equal to the father's force. Since the child is accelerating north, the father must be pushing with a force of <u>40 Newtons north</u>.

11. Since it takes more than 25 Newtons to get the object moving, <u>the static frictional force is 25 Newtons east</u>. Once it is moving, however, it accelerates at 0.1 meters per second^2. This means the total force on the object is:

$$\text{total force} = (\text{mass}) \cdot (\text{acceleration})$$

$$\text{total force} = (15 \text{ kg}) \cdot \left(0.10 \frac{\text{m}}{\text{sec}^2} \right) = 1.5 \text{ Newtons}$$

This force is the combination of the applied force (20 Newtons) and the kinetic frictional force (we know to use the kinetic frictional force because it is moving). Since the kinetic frictional force opposes motion, it is opposite of the applied force. Thus, the total force is the applied force minus the frictional force.

$$20 \text{ Newtons} - \text{kinetic frictional force} = 1.5 \text{ Newtons}$$

Thus, in order for the total force to be 1.5 Newtons, the kinetic frictional force must be <u>18.5 Newtons east</u>.

12. The first part is easy. If the static frictional force is 700 Newtons, <u>the worker must apply more than 700 Newtons of force to get the box moving</u>. To accelerate the box once it is moving, the total force must be:

$$\text{total force} = (\text{mass}) \cdot (\text{acceleration})$$

$$\text{total force} = (500\,\text{kg}) \cdot \left(0.10\,\frac{\text{m}}{\text{sec}^2} \right) = 50\,\text{Newtons}$$

This total force is made up of the worker's force minus the kinetic frictional force. We were told the kinetic frictional force is 220 Newtons, so we can say:

$$\text{worker's force} - 220\,\text{Newtons} = 50\,\text{Newtons}$$

The worker's force, then, must be <u>270 Newtons south</u>.

13. Static friction keeps objects from moving. If the gardener had to exert slightly more than 100 Newtons of force to get the rock moving, <u>the static frictional force is 100 Newtons</u>. Once it got moving, the gardener keeps it moving at a constant velocity eastward. This tells us that the acceleration is zero, which means the total force on the rock is zero. Thus, the gardener applies enough force to overcome the kinetic frictional force, but no more. <u>The kinetic frictional force, then, must be 45 Newtons to the west.</u>

14. The total force on the rock can be calculated from the mass and acceleration:

$$\text{total force} = (\text{mass}) \cdot (\text{acceleration})$$

$$\text{total force} = (710\,\text{kg}) \cdot \left(0.20\,\frac{\text{m}}{\text{sec}^2} \right) = 142\,\text{Newtons}$$

Now what is this force made of? Well, one man is pushing east with 156 Newtons, and the other is pushing east at 220 Newtons. Since those forces are in the same direction, they add. Friction is there as well, however, and it opposes the motion. Thus, it subtracts.

$$156\,\text{Newtons} + 220\,\text{Newtons} - \text{kinetic frictional force} = 142\,\text{Newtons}$$

When we add 156 and 220, we get 376.

$$376\,\text{Newtons} - \text{kinetic frictional force} = 142\,\text{Newtons}$$

So, the kinetic frictional force is equal to whatever number leaves 142 when subtracted from 376. That's 234. Thus, the kinetic frictional force must be <u>234 Newtons west</u>.

15. <u>The equal and opposite force is exerted by the doghouse on the child.</u>

16. <u>The player exerts a force on the ball because the ball's velocity changed. This means there was an acceleration, which means a force was exerted on the ball. The equal and opposite force is exerted by the ball on the player and is evidenced by the pain that the player feels when he catches the ball.</u>

17. The wall exerts a force of <u>20 Newtons west</u>, because it is equal and opposite of the man's force.

SOLUTIONS TO THE MODULE #11 STUDY GUIDE

1. The four fundamental forces are <u>the gravitational force, the weak force, the strong force, and the electromagnetic force.</u> The electromagnetic force and the weak force are really different facets of the same force. Thus, some say there are only 3 fundamental forces in creation.

2. <u>The weakest force is the gravitational force. The strongest one is the strong force.</u>

3. The three principles are:

    1. <u>All objects with mass are attracted to one another by the gravitational force.</u>

    2. <u>The gravitational force between two masses is directly proportional to the mass of each object.</u>

    3. <u>The gravitational force between two masses is inversely proportional to the square of the distance between those two objects.</u>

4. When the 10 kg mass is replaced by a 20 kg mass, the mass doubled. This doubles the gravitational force. The 6 kg mass is also doubled to 12 kg. This once again doubles the gravitational force. Thus, the total change is 2x2 = 4. <u>The new gravitational force, then, is 4 times larger than the old one.</u>

5. The only difference between the first and second situation is the distance between the objects was *multiplied by* 4. The gravitational force decreases when the distance between the objects increases, and it decreases according to the square of that increase. Thus, the force is *divided by* 4^2, which is 16. <u>The new gravitational force, then, is 16 times smaller than the old one.</u>

6. When the 1 kg mass is replaced by a 5 kg mass, the mass was multiplied by 5. This multiplies the gravitational force by 5. The 2 kg mass is doubled to 4 kg. This doubles the gravitational force. The distance between the objects was *divided by* 3. The gravitational force increases with decreasing distance between the objects. The increase goes as the square of the change in the distance. Thus, the force is *multiplied by* 3^2, which is 9. Therefore, the total change is 5x2x9 = 90. <u>The new gravitational force, then, is 90 times larger than the old one.</u>

7. <u>The equal and opposite force is the gravitational force that Venus exerts on the sun.</u>

8. <u>Centripetal force is required for circular motion.</u>

<u>Centripetal force</u> – The force necessary to make an object move in a circle. It is directed perpendicular to the velocity of the object, which means it points toward the center of the circle.

9. The three principles are:

    1. <u>Circular motion requires centripetal force.</u>

    2. <u>The larger the centripetal force, the faster an object travels in a circle of a given size.</u>

    3. <u>At a given speed, the larger the centripetal force, the smaller the circle.</u>

10. The wire is exerting the centripetal force. Thus, the force it exerts will change as needed by the principles listed in question #9.

a. The wire will need to exert <u>more force</u> because objects that travel fast need large centripetal force.

b. The wire will exert <u>less force</u>, because the larger the centripetal force, the smaller the circle. Thus, a larger circle needs less centripetal force (for the same speed).

11. <u>It is a myth. There is no such force.</u> The supposed effect of centrifugal force is just Newton's First Law.

12. Traveling from "A" to "B" tells us that it is traveling clockwise. Its velocity is straight, however. Since it is traveling at a constant speed, the only force is centripetal, which always points to the center of the circle.

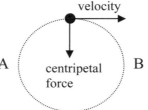

13. Inner Planets – <u>Mercury, Venus, earth, and Mars</u>

Outer Planets – <u>Jupiter, Saturn, Uranus, Neptune</u>

14. <u>Mercury, Venus, earth, Mars, Jupiter, Saturn, Uranus, Neptune</u>

15. <u>Saturn, Uranus, Jupiter, and Neptune</u>

16. <u>Most of the asteroids are between the orbits of Mars and Jupiter.</u> The student could also say that they are in the asteroid belt.

17. <u>Perturbations in its orbit</u> cause an asteroid to become a meteor.

18. A comet is made of <u>a nucleus, a coma, and a tail. The nucleus is always present.</u>

19. All three parts of a comet are present <u>when the comet is close to the sun.</u>

20. <u>Comet orbits are elliptical.</u>

21. Physicists think that short-period comets come from a mass of objects called the <u>Kuiper belt.</u>

22. According to General Relativity, <u>gravity is caused by the fact that objects with mass bend space and time</u>.

23. According to the graviton theory, <u>gravity is caused by the exchange of particles called "gravitons."</u>

SOLUTIONS TO THE MODULE #12 STUDY GUIDE

1 a. <u>Photon</u> – A small "package" of light that acts like a particle

b. <u>Charging by conduction</u> – Charging an object by allowing it to come into contact with an object that already has an electrical charge

c. <u>Charging by induction</u> – Charging an object without direct contact between the object and a charge

d. <u>Electrical current</u> – The amount of charge that travels past a fixed point in an electric circuit each second

e. <u>Conventional current</u> – Current that flows from the positive side of the battery to the negative side. This is the way current is drawn in circuit diagrams, even though it is wrong.

f. <u>Resistance</u> – The ability of a material to impede the flow of charge

g. <u>Open circuit</u> – A circuit that does not have a complete connection between the two sides of the power source. As a result, current does not flow.

2. Like charges repel one another and will thus exert forces pushing the other directly away. Opposite charges attract and will therefore exert forces pulling the other directly in.

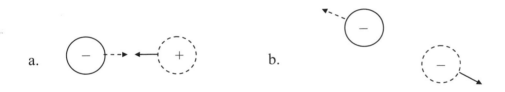

3. The electromagnetic force is inversely proportional to the square of the distance between the objects. Thus, if the distance is multiplied by 3, <u>the force is divided by 9</u>. Since the poles are opposite, <u>it is an attractive force.</u>

4. The electromagnetic force is directly proportional to the charge. When the first charge is doubled, the force is doubled. Since the second charge is left the same, there is no change with respect to that charge. The force varies inversely with the square of the distance between objects. Thus, if the distance is divided by 2, the force is multiplied by 4. The total change, then, is 2 x 4 = 8. <u>The new force is 8 times stronger than the old one.</u>

5. <u>The exchange of photons causes the electromagnetic force.</u>

6. <u>Charged particles do not glow because the photons they emit are not visible to you and me.</u> Under the right conditions, however, charged particles can emit visible light. At those times, you could say that the charged particles do "glow."

7. Charging by induction results in a charge opposite of the rod. Thus, <u>the object will be negatively charged</u>.

8. Charging by conduction results in the same charge as the rod. Thus, <u>the object will be positively charged</u>.

9. Voltage tells us how hard the electricity source pushes on the electrons. This means the larger the voltage, the higher the energy of each electron. Thus, <u>each electron has high energy</u>. Current refers to how many electrons flow through the circuit. Thus, <u>few electrons flow through the circuit</u>. Even though there are few electrons, they each have high energy. Thus, <u>the circuit could be dangerous</u>.

10. <u>A circuit is reasonably safe when both the voltage and the current are low</u>. Please realize that "low" is a relative term. A low voltage is 9 volts or less. A low current is 0.001 Amps or less.

11. Conventional current flows from the positive side of a battery to the negative side.

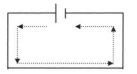

12. <u>Conventional current assumes that electricity is the flow of positive charges</u>. We know, however, that electricity is the flow of electrons, which are negative.

13. The longer the wire, the more chances the electrons have for colliding with atoms within the wire. Thus, <u>the longer wire has more resistance</u>.

14. The wider, or thicker, the wire, the more the electrons can spread out. This means there is less likelihood of electrons colliding with atoms in the wire. Thus, the thicker wire will have less resistance. For the same amount of current, less resistance means less heat. Thus, <u>the thin wire will get hotter</u>. This is one of the causes of house fires. A person uses too thin an extension cord and tries to allow it to run too many devices. This draws too much current for the thin wire, heating it up to the point that it causes a fire.

15. In circuit (a), the open switch makes it impossible for any current to flow. Thus, the light bulb won't glow. <u>In circuit (b), the light bulb glows</u> because the open switch is parallel to the light bulb. Thus, current can still flow through the bulb.

16. <u>The bulbs are wired in a series circuit</u>, because a burnt-out bulb acts like an open switch. If the open switch turns off the bulbs, it is wired in series with the other bulbs.

17. <u>In a permanent magnet, the flow of charged particles is the motion of the electrons in its atoms</u>.

18. As far as we know, <u>this is not possible</u>. Magnets must always have both a north and south pole.

19. <u>Yes</u>, it is possible. If the material responds strongly enough to a magnet, you can align its atoms and make it a magnet.

20. <u>If a material is not magnetic, its atoms cannot be aligned</u>. As a result, the flow of electrons is random, and the material cannot respond to a magnet.

SOLUTIONS TO THE MODULE #13 STUDY GUIDE

1. a. <u>Model</u> – A schematic description of a system that accounts for its known properties

b. <u>Nucleus</u> – The center of an atom, containing the protons and neutrons

c. <u>Atomic number</u> – The number of protons in an atom

d. <u>Mass number</u> – The sum of the numbers of neutrons and protons in the nucleus of an atom

e. <u>Isotopes</u> – Atoms with the same number of protons but different numbers of neutrons

f. <u>Element</u> – A collection of atoms that all have the same number of protons

g. <u>Radioactive isotope</u> – An atom with a nucleus that is not stable

h. <u>Half-life</u> – The time it takes for half of the original sample of a radioactive isotope to decay

2. The three constituents of the atom are the proton, neutron and electron. The electrons are significantly less massive than the other two, and the neutron is just slightly more massive than the proton. Thus, the order is <u>electron, proton, neutron</u>.

3. <u>The nuclear force</u> holds the protons and neutrons in the nucleus. <u>This force is caused by the exchange of pions between protons and/or neutrons</u>. The student could also call it the strong force, since the nuclear force is just another manifestation of the strong force.

4. <u>The electromagnetic force</u> (or electroweak force) holds the electrons in orbit. They stay in orbit because they are attracted to the oppositely charged protons.

5. <u>An atom is mostly empty space</u>.

6. The atomic number is defined as the number of protons in an atom. Atoms have the same number of electrons as they have protons. Thus, this atom has <u>34 electrons and 34 protons</u>. In order to get the symbol, we just look at the chart. The chart tells us that atoms with atomic number of 34 are symbolized with <u>Se</u>.

7. a. Since the chemical symbol is Ne, we can use the chart to learn that the atom has <u>10 protons</u>. This tells us there are also <u>10 electrons</u>. The mass number is the sum of protons and neutrons in the nucleus. Thus, there are also <u>10 neutrons</u>.

b. Since the chemical symbol is Fe, we can use the chart to learn that the atom has <u>26 protons</u>. This tells us there are also <u>26 electrons</u>. The mass number is the sum of protons and neutrons in the nucleus. Thus, there are 30 <u>neutrons</u>.

c. Since the chemical symbol is La, we can use the chart to learn that the atom has <u>57 protons</u>. This tells us there are also <u>57 electrons</u>. The mass number is the sum of protons and neutrons in the nucleus. Thus, there are <u>82 neutrons</u>.

d. Since the chemical symbol is Mg, we can use the chart to learn that the atom has 12 protons. This tells us there are also 12 electrons. The mass number is the sum of protons and neutrons in the nucleus. Thus, there are 12 neutrons.

8. In order to be isotopes, the two atoms must have the same number of protons. Thus, the second atom also has 18 protons.

9. Isotopes must all have the same number of protons but different numbers of neutrons. Since the chemical symbol tells you how many protons that an atom has, in the end, only atoms with the same chemical symbol can be isotopes. In order to be isotopes, then, the atoms must have the same chemical symbol but different mass numbers. Thus, ^{112}Sn, ^{124}Sn, and ^{120}Sn are isotopes.

10. All atoms symbolized with "O" have 8 protons according to the chart. This also means there are 8 electrons. Two of them can go into the first Bohr orbit, but we will have to put the remaining 6 in the second Bohr orbit. That's fine, because the second Bohr orbit can hold up to 8 electrons. The mass number indicates that there are also 8 neutrons:

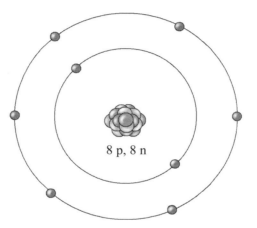

8 p, 8 n

11. All atoms symbolized with "Mg" have 12 protons according to the chart. This also means there are 12 electrons. Two of them can go into the first Bohr orbit, and 8 can go in the second Bohr orbit. We will have to put the remaining 2 in the third Bohr orbit. That's fine, because the third Bohr orbit can hold up to 18 electrons. The mass number indicates that there are 13 neutrons:

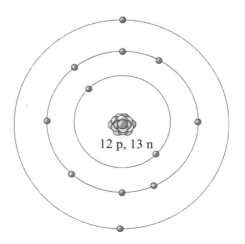

12 p, 13 n

12. All uranium atoms, regardless of their mass number, have 92 protons and 92 electrons. That's what the periodic chart tells us for any element symbolized with "U." The first 2 electrons will go in the first Bohr orbit. The next 8 will go in the second Bohr orbit. The next 18 will go in the third Bohr orbit, and the next 32 will go in the fourth Bohr orbit. That makes 60 electrons. The remaining 32 will all fit in the fifth Bohr orbit because it can hold up to 50 electrons. Thus, the largest Bohr orbit is the fifth one, and there are 32 electrons in it.

13. The strong nuclear force is governed by the exchange of pions. Since pions have a very short lifetime, the strong nuclear force can only act over very tiny distances.

14. a. ^{98}Tc has 43 protons according to the chart. This means there must 55 neutrons. In beta decay, a neutron turns into a proton. This will result in an atom with 44 protons and 54 neutrons, or ^{98}Ru.

b. ^{125}I has 53 protons according to the chart. This means there must be 72 neutrons. In beta decay, a neutron turns into a proton. This will result in an atom with 54 protons and 71 neutrons, or ^{125}Xe.

15. a. ^{212}Bi has 83 protons according to the chart. This means there must be 129 neutrons. In alpha decay, the nucleus loses 2 protons and 2 neutrons. This will result in an atom with 81 protons and 127 neutrons, or ^{208}Tl.

b. ^{224}Ra has 88 protons according to the chart. This means there must be 136 neutrons. In alpha decay, the nucleus loses 2 protons and 2 neutrons. This will result in an atom with 86 protons and 134 neutrons, or ^{220}Rn.

16. Only gamma decay does not affect the number of neutrons and protons in a radioactive isotope.

17. In 1,600 years, the 10 grams will be cut in half to 5 grams. In the next 1,600 years, those 5 grams will be cut in half to 2.5 grams. That's a total of 3,200 years, so the answer is 2.5 grams.

18. In one hour, the ^{11}C will have passed through three half-lives. During the first half-life, the 1 gram sample will be reduced to 0.5 grams. During the next half-life, those 0.5 grams will be reduced to 0.25 grams. In the final half-life, those 0.25 grams will be reduced to 0.125 grams.

19. Radioactive dating is usually unreliable because assumptions must be made as to the original condition of the object. These assumptions are usually erroneous.

20. Alpha particles pass through the least amount of matter before stopping, beta particles are next, and gamma rays pass through the most matter before stopping.

SOLUTIONS TO THE MODULE #14 STUDY GUIDE

1. a. <u>Transverse wave</u> – A wave with a direction of propagation that is perpendicular to its direction of oscillation

b. <u>Longitudinal wave</u> – A wave with a direction of propagation that is parallel to its direction of oscillation

c. <u>Supersonic speed</u> – Any speed that is faster than the speed of sound in the substance of interest

d. <u>Sonic boom</u> – The sound produced as a result of an object traveling at or above Mach 1

e. <u>Pitch</u> – An indication of how high or low a sound is, which is primarily determined by the frequency of the sound wave

2. <u>The engineers need to adjust the electronics to emit sound waves with a shorter wavelength.</u> Remember, wavelength and frequency are inversely proportional. If the engineers want higher pitch, they want higher frequency, which means they want a shorter wavelength.

3. To determine the speed of sound, we use Equation (14.2):

$$v = (331.5 + 0.6 \cdot 30) \frac{m}{sec}$$

$$v = (331.5 + 18) \frac{m}{sec}$$

$$\underline{v = 349.5 \frac{m}{sec}}$$

4. To determine a wave's frequency, we use Equation (14.1):

$$f = \frac{v}{\lambda}$$

$$f = \frac{349.5 \frac{m}{sec}}{0.5 \; m} = 699 \; \frac{1}{sec}$$

The frequency is <u>699 Hz</u>.

5. Infrasonic waves have frequencies less than 20 Hz (the lowest frequencies that human ears can hear). Ultrasonic waves have frequencies of more than 20,000 Hz (the highest frequencies that human ears can hear). Sonic waves have frequencies between 20 and 20,000 Hz. To determine whether a

wave is sonic, infrasonic, or ultrasonic, then, we must determine its frequency. That's what we can do with Equation (14.1):

$$f = \frac{v}{\lambda} = \frac{345 \ \frac{\cancel{m}}{sec}}{500 \ \cancel{m}} = 0.69 \ \frac{1}{sec}$$

Since the frequency is less than 20 Hz, this is an <u>infrasonic wave</u>.

6. <u>The physicist will not be able to hear the alarm because, without air, the sound waves from the alarm have nothing through which to travel. Thus, they cannot make waves. As a result, there is no sound.</u>

7. <u>Sound waves are longitudinal waves.</u>

8. To determine the distance, we will use the time difference between the lightning flash and the sound. We will assume that the light from the lightning reaches our eyes essentially at the same instant as the lightning was formed. Thus, the time it takes for the sound to travel to you will determine the distance. First, then, we need to know the speed of sound:

$$v = (331.5 + 0.6 \cdot 13) \ \frac{m}{sec}$$

$$v = (331.5 + 7.8) \ \frac{m}{sec}$$

$$v = 339.3 \ \frac{m}{sec}$$

Now we can use Equation (14.3):

$$\text{distance traveled} = (\text{speed}) \times (\text{time traveled})$$

$$\text{distance traveled} = (339.3 \frac{m}{\cancel{sec}}) \times (2.3 \ \cancel{sec}) = \underline{780.39 \ m}$$

9. Sound travels faster in solids than it does in gases or liquids. Thus, <u>the sound travels faster in the wall</u>.

10. Remember, when a wave strikes an obstacle, part of the wave is reflected and part of it is transmitted through the obstacle. Thus, only a portion of the wave actually starts traveling through the wall. This means <u>the amplitude of the wave will be smaller</u>, because only part of the wave is present in the wall.

11. To determine the speed of the jet, we first have to determine the speed of sound. After all, Mach 2.5 means 2.5 times the speed of sound. Thus, we need to know the speed of sound in order to determine the speed of the jet.

$$v = (331.5 + 0.6 \cdot T) \frac{m}{sec}$$

$$v = (331.5 + 0.6 \cdot 1) \frac{m}{sec}$$

$$v = (331.5 + 0.6) \frac{m}{sec} = 332.1 \frac{m}{sec}$$

Since sound travels at 332.1 m/sec, Mach 2.5 is 2.5 x (332.1 m/sec), or 830.25 m/sec.

12. This is much like the previous problem. To determine the Mach number, we need to first determine how quickly sound travels in that air:

$$v = (331.5 + 0.6 \cdot T) \frac{m}{sec}$$

$$v = (331.5 + 0.6 \cdot 0) \frac{m}{sec} = 331.5 \frac{m}{sec}$$

To determine the Mach number, then, we can just divide the speed of the jet by the speed of sound. This will tell us how many times faster than sound the jet is traveling:

$$464.1 \div 331.5 = 1.4$$

This means that the jet is traveling at Mach 1.4.

13. When a jet travels at Mach 1 or higher, it creates a shock wave of air that causes a very loud boom. This boom can damage people's ears and buildings. Thus, sonic booms must be avoided when people or buildings are nearby.

14. Since the string becomes shorter, the wavelength will be smaller. This will result in a higher frequency. Thus, the pitch will increase.

15. The wavelength, frequency, and speed will all be the same. After all, the pitch is determined by the frequency, which, in turn, determines the wavelength. The speed depends only on the temperature. The amplitudes of the waves will be different, however, because amplitude determines loudness.

16. As the car travels away from you, the sound waves that are produced by the horn get farther and farther apart. This makes the wavelength seem longer to your ears, which will result in a lower frequency. Thus, the horn's pitch is lower when it is moving away from you.

17. As you travel towards the police car, you will encounter the compressions of the sound waves faster than when you stand still. As a result, the wavelength will seem shorter to you, which will produce higher frequencies. This means that the pitch will get <u>higher</u>.

18. The bel scale states that every bel unit corresponds to a factor of ten increase in the intensity of the sound waves. Thus, we need to determine how many bel units the sound of the traffic is, as compared to the sound of your voice:

$$\frac{80 \text{ decibels}}{1} \times \frac{0.1 \text{ bel}}{1 \text{ decibel}} = 8 \text{ bels}$$

$$\frac{100 \text{ decibels}}{1} \times \frac{0.1 \text{ bel}}{1 \text{ decibel}} = 10 \text{ bels}$$

Since the traffic is 2 bels louder than your voice, the increase in sound wave intensity is 2 factors of ten higher. Thus, the traffic has sound waves with intensities that are 10 x 10 = <u>100 times larger than the intensities of sound waves from your voice.</u>

19. The bel scale states that every bel unit corresponds to a factor of ten increase in the intensity of the sound waves. Thus, we need to determine how many bel units were fed into the amplifier:

$$\frac{30 \text{ decibels}}{1} \times \frac{0.1 \text{ bel}}{1 \text{ decibel}} = 3 \text{ bels}$$

If the amplifier magnifies the intensities of the waves by a factor of 1,000, that's the same as 10 x 10 x 10. Thus, the sound coming out of the amplifier will be 3 bels larger, as each factor of ten represents one increase in the bel level. Thus, the sound coming out will be 6 bels, which is the same as <u>60 decibels</u>.

SOLUTIONS TO THE MODULE #15 STUDY GUIDE

1. a. <u>Electromagnetic wave</u> – A transverse wave composed of an oscillating electric field and a magnetic field that oscillates perpendicular to the electric field

b. <u>The Law of Reflection</u> – The angle of reflection equals the angle of incidence.

2. <u>The wave theory of light views light as two transverse waves, one made of an oscillating magnetic field and the other an oscillating electric field. The particle theory of light views a ray of light as a beam of individual particles called photons. The quantum-mechanical theory says that light is both a particle and a wave. It is made up of individual packets that behave like particles, but the packet is actually made up of a wave.</u>

3. <u>Light waves oscillate a magnetic field and an electric field.</u> Each one oscillates perpendicular to the other, as well as perpendicular to the direction of travel.

4. Einstein's Special Theory of Relativity says that <u>nothing with mass can travel faster than the speed of light</u>.

5. Unlike sound, light travels slower in liquids than gases. Thus, <u>the light's speed increased once it left the water</u>.

6. The acronym ROY G. BIV allows us to remember the relative size of the colors' wavelengths. Red is longest and violet is shortest. Thus, in terms of increasing wavelength, the colors are: <u>violet, green, yellow, and orange</u>.

7. Wavelength and frequency are inversely proportional. Thus, in terms of increasing frequency, it is the reverse of the previous order: <u>orange, yellow, green, and violet</u>.

8. Radio waves have wavelengths longer than visible light while X-rays have shorter wavelengths. This comes from Figure 15.3. Just like the visible light colors, you need not memorize any values for the wavelengths, but of the major categories in the figure, you need to know their relative wavelengths. Since frequency and wavelength are inversely proportional, <u>radio waves have lower frequencies than visible light, while X-rays have higher frequencies</u>.

9. <u>Infrared light is not visible</u>. Thus, even though human bodies constantly emit infrared light, we cannot see that happening. There are special devices you can get that do, indeed, detect the infrared light that the human body emits. This allows you to see living organisms and other hot objects, even in the darkest of nights.

10. By the Law of Reflection, <u>the reflected light also makes a 15 degree angle relative to that line</u>.

11. <u>Yes</u>. Remember, to see a part of his body, light must be able to travel from that part of his body, reflect off the mirror, and hit his eye. His brain will then extend that line backwards, forming an image in the mirror. The only constraint is that the light that strikes the mirror must obey the Law of Reflection:

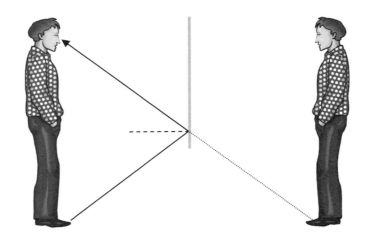

Had the mirror been much smaller, there would have been no way light could travel from his foot and hit his eyes and also obey the Law of Reflection:

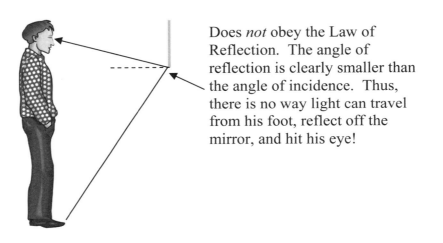

Does *not* obey the Law of Reflection. The angle of reflection is clearly smaller than the angle of incidence. Thus, there is no way light can travel from his foot, reflect off the mirror, and hit his eye!

12. <u>The light ray can be reflected or refracted.</u>

13. When light travels from a substance in which it moves quickly to a substance in which it moves slowly, the light bends towards the perpendicular. Since light moves faster in air than in glass, <u>the light will bend towards the perpendicular line.</u>

14. In order for you to see the objects underwater, light must travel from the object to your eyes. Thus, the light must travel out of the water and into the air. <u>When light travels from water to air, it bends. This causes your mind to form a false image of the object in a slightly different location.</u>

15. <u>There must be water droplets suspended in the air, the sun must be shining on them from behind you, and the sun must be at a certain angle (or height in the sky).</u> The water droplets cause the refraction. In order to separate the colors enough to see them, however, the light must be refracted, reflected, and refracted again (see Figure 15.6). To do that, light must enter the water droplet on the side from which you are viewing it.

16. <u>A converging lens causes light rays to bend so that they converge to a single point. Diverging lenses cause light rays to bend away from each other.</u>

17. The function of a lens depends on its shape. <u>Lens (a) is a converging lens and lens (b) is a diverging lens.</u>

18. <u>The eye focuses light by changing the shape of its lens. A camera focuses light by moving the position of the lens.</u> The eye's method is faster and much more precise.

19. If your red cone cells no longer worked, your brain would think that you never saw red light. When the white light reflected off of the paper and hit your eyes, then, your green cone cells would send signals to your brain, as would your blue cone cells. As a result, <u>the white paper would appear to be blue-green or cyan.</u> If you looked at a red piece of paper, it is red because it only reflects red light. Thus, it would send only red light to your eyes. You cannot see red light, however, <u>so the red paper would appear to be black</u>. In actual fact, it would probably appear gray, because the other cells in your eyes (rods) detect the presence of light without detecting color. Thus, your brain would receive signals from the rods telling it that light was there, but there would be no cone cell signals to indicate the color. The brain would be a bit confused, but it would probably form a grayish image, as if you were seeing only in black and white.

20. In order to look violet, it must absorb all colors except violet. <u>Thus, it absorbs red, yellow, orange, green, blue, and indigo light.</u>

21. Since cyan absorbs all colors except blue and green, it will absorb any red light shone on it. As a result, no light will make it to your eyes. <u>In red light, then, the paper would look black.</u> When you shined green light on it, the green light would be reflected. There would be no blue to mix with it, though. <u>In green light, the paper would look green.</u>

SOLUTIONS TO THE MODULE #16 STUDY GUIDE

1. a. <u>Nuclear fusion</u> – The process by which two or more small nuclei fuse to make a bigger nucleus

b. <u>Nuclear fission</u> – The process by which a large nucleus is split into smaller nuclei

c. <u>Critical mass</u> – The amount of isotope necessary to sustain a chain reaction

d. <u>Absolute magnitude</u> – The brightness of a star, corrected for distance, on a scale of -8 to +19. The *smaller* the number, the *brighter* the star.

e. <u>Apparent magnitude</u> – The brightness of a star as seen in the night sky. The *smaller* the number, the *brighter* the star.

f. <u>Light year</u> – The distance light could travel along a straight line in one year

g. <u>Galaxy</u> – A large ensemble of stars, all interacting through the gravitational force and orbiting around a common center

2. Starting on the inside, the sun is divided into <u>the core, the radiative zone, the convection zone, and the photosphere.</u>

3. <u>The sun gets its power from nuclear fusion that occurs in the core.</u>

4. We see <u>the photosphere.</u>

5. Since a large nucleus split into two smaller nuclei (and a few neutrons), this is <u>nuclear fission.</u>

6. Since two small nuclei became a bigger nucleus (plus a neutron), this is <u>nuclear fusion.</u>

7. In both processes, mass is converted into energy. Thus, <u>the mass of the starting materials is larger than the mass of the materials the process makes.</u>

8. It is impossible for a nuclear power plant to experience a nuclear explosion because <u>a power plant does not have significantly more than the critical mass of the large nucleus that is breaking apart.</u>

9. <u>Nuclear fusion is a better means of producing energy because there are no radioactive byproducts, there is no chance of meltdown, and the starting materials are cheap.</u>

10. <u>We cannot use nuclear fusion yet because we cannot master the technology to make it economically feasible.</u>

11. To classify a star, you find where its magnitude and spectral letter put it on the H-R Diagram. Taking the values given for the stars in this problem, you can come up with the following H-R Diagram:

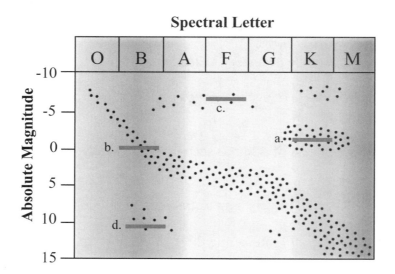

Thus:

a. <u>Red giant</u>   b. <u>Main sequence</u>   c. <u>Supergiant</u>   d. <u>White dwarf</u>

12. Our sun is a main sequence star. Thus, <u>star (b) is most like our sun</u>.

13. In general, supergiants are the largest, red giants are next, main sequence stars are next, and white dwarfs are the smallest. Thus, in terms of *increasing* size, it is <u>(d), (b), (a), and (c)</u>.

14. Brightness is given by magnitude. The *smaller* the magnitude, however, the brighter the star. Thus, in terms of *increasing* brightness, it is <u>(d), (b), (a), and (c)</u>.

15. The farther to the right on the H-R Diagram, the cooler the star. Thus, <u>(a) is the coolest</u>.

16. <u>All three of these are variable star types. Thus, their brightness changes radically with time.</u>

17. <u>The big difference between these star types is lifetime. Pulsating stars last a long time, supernovas exist very briefly, and novas are somewhere in between.</u>

18. <u>The crab nebula was most likely formed by a supernova.</u>

19. <u>The two methods are the parallax method and the apparent magnitude method. The parallax method is exact, but the apparent magnitude method can be used to measure longer distances.</u>

20. Cepheid variables are important for measuring long distances because they seem to have a relationship between their period and their magnitude. That allows them to be used in the apparent magnitude method for measuring long distances in the universe.

21. The four galaxy types are spiral, lenticular, elliptical, and irregular. The Milky Way is a spiral galaxy.

22. Stars group together to form galaxies, which group together to form groups, which group together to form clusters, some of which group together to form superclusters.

23. The earth's solar system belongs in the Milky Way, which belongs to the Local Group, which belongs to the Virgo Cluster.

24. Most astronomers believe that the universe is expanding because the light from nearly every galaxy experiences a red shift before it reaches the earth, and the red shift increases the farther the galaxy is from the earth.

25. Yes, the geometry of the universal expansion makes a great deal of difference. The theories that can be developed for the formation of the universe depend on that initial assumption.

Answers to the Module Summaries in Appendix B

Not all of the blanks have to be filled in with exactly the phrases used here. As long as the general message of each paragraph is the same, that's fine.

ANSWERS TO THE SUMMARY OF MODULE #1

1. Atoms are the <u>smallest chemical</u> unit of matter. They are so <u>small</u> that you cannot see them. Images of atoms produced by scanning tunneling electron microscopes are not pictures, but are the result of computer <u>calculations</u>. Such images are part of the large amount of <u>indirect</u> evidence that indicates atoms exist. When two or more atoms link together, they form a <u>molecule</u>, which has its own unique properties.

2. When a substance is made of billions and billions of the same atom, it is called an <u>element</u>, while substances made up of billions and billions of the same molecule are called <u>compounds</u>. Some substances, called <u>mixtures</u>, are made of more than one kind of atom or molecule. In Experiment 1.1, you watched the molecule known as <u>water</u> break down into two elements: <u>hydrogen</u> and <u>oxygen</u>. In addition, you saw the <u>copper</u> in the wire (an element) react with <u>baking soda</u> molecules to make copper hydroxycarbonate.

3. When making measurements, the <u>units</u> we report are just as important as the numbers. The base metric unit for mass is the <u>gram</u>, while the base English unit for mass is the <u>slug</u>. The base metric unit for weight is the <u>Newton</u>, while the base English unit for weight is the <u>pound</u>. Although the weight of an object varies depending on gravity, the <u>mass</u> does not.

4. The base metric unit for length is the <u>meter</u>, and the base English unit for length is the <u>foot</u>. The base metric unit for volume is the <u>liter</u>, and the base English unit for volume is the <u>gallon</u>.

5. In the metric system, prefixes govern the <u>size</u> of a unit. The "milli" prefix means <u>0.001</u>, while the "centi" prefix means 0.01. The "kilo" prefix means <u>1,000</u>. Thus, an object with a mass of 1 centigram has <u>less</u> mass than an object with a mass of 1 kilogram.

6. The metric unit for temperature is degrees <u>Celsius</u>. This unit uses no prefixes. In this system, water freezes at <u>0 degrees Celsius</u> and boils at <u>100 degrees Celsius</u>.

7. In the Old Testament, a measurement unit for length called the <u>cubit</u> was used.

8. Concentration is the <u>quantity</u> of a substance within a certain <u>volume</u>. At certain concentrations, chemicals <u>behave in one way</u>. At other concentrations, those same chemicals can <u>behave in a different way</u>. Vitamins, for example, are <u>very good</u> for your body at low concentrations, but they become <u>toxic</u> at high concentrations. In Experiment 1.3, the more concentrated the vinegar, the <u>faster</u> the antacid tablet disappeared.

9. In this course, we do conversions using the <u>factor-label</u> method, in which the measurement you want to convert is multiplied by a fraction that contains both the original unit and the unit you want to convert to.

10. $\dfrac{34.5 \, \cancel{kL}}{1} \times \dfrac{1{,}000 \, L}{1 \, \cancel{kL}} = \underline{34{,}500 \, L}$

11. $\dfrac{0.00045 \, g}{1} \times \dfrac{1 \, mg}{0.001 \, g} = \underline{0.45 \, mg}$

12. To properly compare these measurements, we need to get them into the same units. I will convert mL to L, although you could just as easily convert L to mL.

$$\dfrac{45 \, \cancel{mL}}{1} \times \dfrac{0.001 \, L}{1 \, \cancel{mL}} = 0.045 \, L$$

Since 45 mL is really equal to 0.045 L, it is the smallest of the two numbers. Thus, the 0.6 L takes up more volume.

13. $\dfrac{13.1 \, g}{1} \times \dfrac{1 \, cg}{0.01 \, g} = \underline{1{,}310 \, cg}$

14. $\dfrac{45.1 \, \cancel{mL}}{1} \times \dfrac{0.001 \, L}{1 \, \cancel{mL}} = \underline{0.0451 \, L}$

15. $\dfrac{13.1 \, \cancel{miles}}{1} \times \dfrac{5{,}280 \, feet}{1 \, \cancel{mile}} = \underline{69{,}168 \, feet}$

16. $\dfrac{16.2 \, \cancel{m}}{1} \times \dfrac{1 \, cm}{0.01 \, \cancel{m}} = \underline{1{,}620 \, cm}$

17. $\dfrac{345.6 \, \cancel{mg}}{1} \times \dfrac{0.001 \, g}{1 \, \cancel{mg}} = \underline{0.3456 \, g}$

18. $\dfrac{100.0 \, \cancel{yards}}{1} \times \dfrac{3 \, feet}{1 \, \cancel{yard}} = \underline{300 \, feet}$

19. $\dfrac{451 \, \cancel{g}}{1} \times \dfrac{1 \, kg}{1{,}000 \, \cancel{g}} = \underline{0.451 \, kg}$

ANSWERS TO THE SUMMARY OF MODULE #2

1. The moisture content of air is called <u>humidity</u>. God has designed you to <u>sweat</u> when you are too warm. This releases water onto your skin, which then evaporates. The process of evaporation requires <u>energy</u>, which is supplied by your skin. As a result, when your sweat evaporates, the net effect is that your skin <u>cools down</u>. When the humidity is high, your sweat does not <u>evaporate</u> as quickly, and as a result, you do not cool down as well. For this reason, many weather reports include a <u>heat index</u>, which is a combination of temperature and humidity.

2. There are two ways of reporting humidity: <u>relative humidity</u> and <u>absolute humidity</u>. If you report the mass of water vapor contained in a certain volume of air, you are reporting the <u>absolute humidity</u>. If you report the ratio of the mass of water vapor in the air at a given temperature to the maximum mass of water vapor the air could hold at that temperature, you are reporting the <u>relative humidity</u>.

3. On a day when the relative humidity is high, water evaporates <u>slowly</u>. On a day when the relative humidity is low, water evaporates more <u>quickly</u>. If the relative humidity is <u>100%</u>, we say that the air is saturated with water.

4. Dry air (air that has all of the <u>water vapor</u> removed) is 78% <u>nitrogen</u>, <u>21%</u> oxygen, and 1% <u>other gases</u>. This is an <u>ideal</u> mixture of gases to support life. The oxygen is necessary in order to allow our bodies to run <u>combustion</u> reactions. Without enough oxygen, our bodies would run out of the <u>energy</u> necessary for life. Too much oxygen in the atmosphere, however can cause <u>health</u> problems in people and significantly increase the number of natural <u>forest fires</u>.

5. The majority of the air we breathe in is nitrogen, and the majority of the air that we breathe out is <u>nitrogen</u>. In addition to nitrogen, the other major gases we exhale are <u>oxygen</u>, <u>water vapor</u>, and <u>carbon dioxide</u>. Of those three gases, we exhale significantly more <u>oxygen</u> than the other two. <u>Carbon dioxide</u> enters the atmosphere as a result of organisms breathing and as the result of fires.

6. Through a process called photosynthesis, plants convert <u>carbon dioxide</u> and water into glucose, which they use for food. A byproduct of this process is <u>oxygen</u>. In addition to allowing plants to manufacture their own food, <u>carbon dioxide</u> helps regulate the temperature of the earth. Through a process referred to as the <u>greenhouse effect</u>, this gas traps heat that radiates from the earth. Without such gases, the average temperature of the earth would be far too <u>low</u> to support life.

7. If the greenhouse effect ran out of control, we would have <u>global warming</u>, a situation in which the average temperature of the earth increased over time. Although this would be bad, the greenhouse effect is <u>good</u>, as it makes life on earth possible. Since carbon dioxide is a gas that participates in the greenhouse effect, the fact that the concentration of carbon dioxide in the atmosphere is <u>rising</u> makes some people fear that <u>global warming</u> may already be happening.

8. The average temperature of the earth has changed <u>little</u> in the past 100 years. There was a small <u>increase</u> in the average global temperature from the late 1800s to the early 1900s, but that was before carbon dioxide levels in the atmosphere <u>increased</u> significantly. Thus, there is <u>little</u> evidence that increasing carbon dioxide concentration causes <u>global warming</u>. In addition, the earth was significantly <u>warmer</u> between the ninth and fourteenth centuries than it is today.

9. Ultraviolet light is <u>harmful</u> to living organisms, because it has enough energy to <u>kill</u> living tissue. Although the sun produces a significant amount of ultraviolet light, most of it is blocked by <u>ozone</u> in the <u>ozone</u> layer. This gas is <u>poisonous</u>, but the <u>ozone</u> layer is high above sea level, where no one is breathing. Even though most ultraviolet light from the sun is blocked, some gets through, and if your skin is exposed to too much of it, you can get a <u>sunburn</u>.

10. When we report concentration in parts per million, we are reporting the number of molecules (or atoms) of a substance in a mixture for every 1 <u>million</u> molecules (or atoms) in that mixture. The concentration of many pollutants is often expressed in parts per million, as their concentrations are very <u>low</u>. Despite what many people think, the concentrations of pollutants in earth's atmosphere have been <u>decreasing</u> for some time. Thus, the air we are breathing today is <u>cleaner</u> than it was 30 years ago.

11. Sulfur oxides are put in the atmosphere when sulfur <u>burns</u>. Sulfur is a <u>contaminant</u> in all <u>fuels</u> we burn. The sulfur content of coal can be <u>reduced</u> in a process called "cleaning." Many industries have devices in their <u>smokestacks</u> that help clean the sulfur oxides out of the mixture of gases that result from burning fuel. These devices are commonly called <u>scrubbers</u>. Although human activity puts sulfur oxides in the atmosphere, there are natural sources as well, one of the most important being <u>volcanoes</u>.

12. Nitrogen oxides are formed when <u>nitrogen</u> burns. This happens at very high <u>temperatures</u>, so engines and power plants are major human-made sources of nitrogen oxides. There are, however, many natural sources of nitrogen oxides as well, such as <u>volcanoes</u>, <u>lightning</u>, and <u>biological decay</u>.

13. While ozone in the ozone layer protects living organisms, ozone is a <u>poison</u>. Thus, <u>ground</u>-level ozone is a pollutant, because people breathe it.

14. <u>Carbon monoxide</u> is a toxic byproduct of incomplete combustion. Unlike carbon dioxide, this gas can be <u>deadly</u>, even at concentrations as low as a few hundred parts per million. Automobiles used to be a major source of this gas, but the introduction of <u>catalytic converters</u> significantly reduced the amount produced in automobiles by converting the gas into <u>carbon dioxide</u>.

15. The U.S. government began issuing standards called <u>CAFE standards</u>, which regulate the average number of miles an automobile can travel on a single gallon of gasoline. Although these standards do <u>reduce</u> automobile-related pollution, they also <u>increase</u> the number of traffic fatalities.

16. When one compares the positive result of an action to the negative result and decides whether or not the positive result was worth the accompanying negative result, we say that the person has done a <u>cost/benefit analysis</u>. Any reasonable discussion of air pollution <u>regulations</u> must include such an analysis in order to ensure that the reduced pollution is worth whatever accompanying costs exist.

17. Remember, we know the relationship between percent and ppm, so we can convert using the factor-label method.

$$\frac{0.03 \; \cancel{\%}}{1} \times \frac{10,000 \; ppm}{1 \cancel{\%}} = 300 \; ppm$$

The concentration of argon in the air is <u>300 ppm</u>.

18. Remember, we know the relationship between ppm and percent. We can therefore just use the factor-label method to figure out the answer.

$$\frac{151 \, \cancel{ppm}}{1} \times \frac{1\%}{10,000 \, \cancel{ppm}} = 0.0151\%$$

A concentration of 151 ppm is equal to <u>0.0151%</u>.

19. $\dfrac{0.091 \, \cancel{ppm}}{1} \times \dfrac{1\%}{10,000 \, \cancel{ppm}} = \underline{0.0000091\%}$

20. $\dfrac{0.011 \, \cancel{\%}}{1} \times \dfrac{10,000 \, ppm}{1 \cancel{\%}} = \underline{110 \, ppm}$

ANSWERS TO THE SUMMARY OF MODULE #3

1. The mass of air surrounding a planet is called its <u>atmosphere</u>. Everything that comes into contact with the mass of air surrounding the earth is subjected to <u>atmospheric pressure</u>, which is, on average, 14.7 pounds per square inch at sea level. Even though this is a lot of pressure, we don't feel it, because <u>equal</u> pressure pushes on you from all sides, even from within.

2. In Experiment 3.1, the cans were filled with steam that turned into <u>liquid water</u> when the cans were put in ice water. The upright can did not crumple, however, because the steam was replaced with <u>air</u> that continued to exert <u>pressure</u> on the inside of the can. The can placed <u>upside down</u> in the water did crumple, however, because <u>air</u> could not replace the steam, so very little <u>pressure</u> was being exerted inside the can.

3. A <u>barometer</u> measures atmospheric pressure. It is composed of a tube with no <u>air</u> inside that is inverted over a pool of <u>liquid</u>, which is usually mercury. Since the <u>atmosphere</u> is pressing down on the pool, and since there is no air exerting <u>pressure</u> inside the tube, <u>liquid</u> is forced up the tube. The <u>height</u> of the liquid in the tube is a measure of the <u>atmospheric pressure</u>.

4. When measuring atmospheric pressure, several units can be used. <u>Pounds per square inch</u> tells you how many pounds are exerted on an 1-inch by 1-inch square. When reported in <u>inches</u>, it tells you the height of a column of mercury in a barometer, in English units. When reported in <u>mmHg</u>, it tells you the height of the column of mercury in metric units. Finally, pressure can also be reported in <u>atm</u>, which tells you the pressure relative to earth's average atmospheric pressure at sea level.

5. The atmosphere can be divided into two general layers. The <u>homosphere</u> is the lower layer, and it contains air that has the same <u>composition</u> as the air at sea level. The upper layer is called the <u>heterosphere</u>, and the mixture of gases in this layer is not <u>uniform</u>. Throughout both layers, however, the total amount of air <u>decreases</u> with increasing altitude.

6. The homosphere is generally divided into three regions. From lowest to highest, they are the <u>troposphere</u>, the <u>stratosphere</u>, and the <u>mesosphere</u>. The first two layers are separated by the <u>tropopause</u>; the second two are separated by the <u>stratopause</u>, and the last layer is separated from the heterosphere by the <u>mesopause</u>. Throughout all of these regions, as well as the heterosphere, the atmospheric pressure continually <u>decreases</u> with increasing altitude.

7. In the troposphere, the temperature steadily <u>decreases</u> with increasing altitude. This is called the temperature <u>gradient</u> of the troposphere. The troposphere is often called earth's <u>weather</u> layer, because it contains almost all of earth's clouds, rain, snow, storms, etc

8. Narrow bands of high-speed winds that circle the earth, blowing from west to east are called <u>jet streams</u>. They are found in the <u>lower</u> portions of the stratosphere and the <u>upper</u> portions of the troposphere. They tend to <u>steer</u> storms and affect which parts of the earth experience high <u>atmospheric pressure</u> or low <u>atmospheric pressure</u>.

9. In the stratosphere, the temperature <u>increases</u> with increasing altitude. This is mostly due to the <u>ozone layer</u>, which is found there. In the mesosphere, the temperature <u>decreases</u> with increasing altitude. When rocks from outer space fall into the mesosphere and burn up, they are called <u>meteors</u>.

10. When energy is transferred as a consequence of temperature differences, we call it <u>heat</u>. When an object gains energy, the <u>molecules (or atoms)</u> that make it up move faster. A <u>thermometer</u> really measures the average speed at which the <u>molecules (or atoms)</u> of a substance are moving. As a result, temperature is a measure of the <u>energy</u> of <u>random motion</u> in a substance's <u>molecules (or atoms)</u>.

11. The "hole" in the ozone layer is actually a seasonal <u>decrease</u> in the concentration of ozone in the ozone layer. It is centered over <u>Antarctica</u>. Human-made substances called <u>CFCs (chlorofluorocarbons)</u> are at least partially to blame. Unlike many chemicals, they are so <u>unreactive</u> that they can survive the trip up to the ozone layer, where they can destroy ozone. Interestingly enough, this same property makes them <u>non-toxic</u> to human beings. They are very efficient chemicals that can be used for <u>firefighting</u>, <u>refrigeration</u>, and <u>surgical sterilizers</u>. Despite their usefulness, their elimination has been called for by the <u>Montreal Protocol</u>. This will, mostly likely, cause an <u>increase</u> in the number of people who die each year.

12. The "hole" in the ozone layer was discovered <u>long before</u> CFCs were widely used. However, since the production of CFCs, the "hole" has gotten "<u>deeper</u>." Although CFCs are heavy, they are lifted up to the ozone layer by the <u>polar vortex</u>. This is why the ozone "hole" is a <u>seasonal</u> phenomenon and why it is centered over <u>Antarctica</u>. While the elimination of CFCs will <u>reduce</u> the depth of the ozone "hole," it will most likely <u>cost</u> more lives than it will save. The ozone "hole" does *not* contribute to <u>global warming</u>. In fact, a reduction in the amount of ozone in the ozone layer will <u>reduce</u> the average temperature of the earth.

13. The heterosphere is generally divided into two layers: the <u>thermosphere</u> and the <u>exosphere</u>. While the <u>thermosphere</u> is lower than the <u>exosphere</u>, they can both be considered a part of outer space. The number of molecules in the <u>thermosphere</u> is so small that a thermometer would read incredibly low temperatures. However, the average <u>energy</u> of the few molecules that are there is very <u>high</u>. The <u>exosphere</u> is composed of those atoms and molecules actually in orbit around the <u>earth</u>. It is difficult to say where the <u>exosphere</u> ends and interplanetary space begins.

14. Between the upper portions of the <u>mesosphere</u> and the lower portions of the <u>thermosphere</u>, there is a region where the atmosphere's gases are ionized. It is called the <u>ionosphere</u>. Atoms are composed of <u>protons</u> (which have positive electrical charge), <u>electrons</u> (which have negative electrical charge), and <u>neutrons</u> (which have no electrical charge). Atoms always have the same number of <u>electrons</u> and <u>protons</u>. This means that overall, atoms have no net <u>electrical charge</u>. When an atom loses (or gains) electrons, there is an <u>imbalance</u> of positive and negative charges, and the atom becomes <u>electrically charged</u>. When this happens, it is no longer an atom, but is instead an <u>ion</u>.

15. The Northern Lights and Southern Lights are examples of <u>auroras</u>. They appear in the night sky as glowing regions of brilliant <u>colors</u> that tend to move over the sky in interesting ways. They are the result of high-energy <u>collisions</u> between ionized particles in the <u>ionosphere</u>.

ANSWERS TO THE SUMMARY OF MODULE #4

1. We can live for as many as two weeks without food, but if we were to go even a few days without water, we would surely die. Indeed, without water, life as we know it simply cannot exist.

2. The use of electricity to break a molecule down into smaller units is called electrolysis. When you use this procedure on water, you produce hydrogen and oxygen. If you measure the volumes of each, you will find twice as much hydrogen as oxygen.

3. Often, experiments produce incorrect results due to experimental error. A good scientist tries to reduce the amount of it in an experiment and does not trust experiments that have a lot of it. When a scientist performs an experiment that seems to lead to a new, interesting conclusion, other scientists in the field look closely at the experiment in an effort to spot experimental errors that the original scientist did not recognize. This process is called peer review. An example of the importance of this process is seen in Drs. Martin Fleischmann and Stanley Pons, who claimed to have discovered cold fusion, a potential source of cheap, unlimited power. They did not submit their experiments to peer review before publicizing their results. As a result, they were embarrassed because other scientists had to publicly announce the experimental errors in their experiments.

4. The chemical symbol of an atom contains either one or two letters. If there are two letters, only the first is capitalized. The letters often come from either the English or Latin name of the atom. A chemical formula tells you the composition of a molecule because the subscripts in the chemical formula tell you how many of each atom is present. If there is no subscript next to an atom's symbol, there is one of those atoms in the molecule. Thus, the molecule $CaCO_3$ has one calcium (Ca) atom, one carbon (C) atom, and three oxygen (O) atoms.

5. Magnesium hydroxide is a chemical often used in antacid tablets. It has one Magnesium (Mg) atom, two oxygen (O) atoms, and two hydrogen (H) atoms. Thus, its chemical formula is MgO_2H_2. The chemical formula of sodium nitrate is $NaNO_3$. This molecule has a total of 5 atoms: 1 sodium (Na) atom, 1 nitrogen (N) atom, and 3 oxygen (O) atoms.

6. Atoms in molecules are linked together with chemical bonds, which are made up of shared electrons. If the atoms in a molecule do not share electrons equally, small charges result within the molecule, and it is called a polar molecule. In Experiment 4.2, the small positive charges on the hydrogen atoms of the water molecules were attracted to the negative charge on the comb. That's what made the stream of water bend towards the comb.

7. When you dissolve a substance in a liquid, we say you have made a solution. When making a solution, you use a solvent to dissolve a solute. When you dissolve sugar in water, for example, water is the solvent, sugar is the solute, and sugar water is the solution.

8. In general, a polar or ionic solute can only dissolve in a polar solvent. A nonpolar solute can only dissolve in a nonpolar solvent. A solute dissolves in a solvent because the molecules of the solvent are attracted to the molecules (or ions) of the solute.

9. In a water molecule, the positive charge on one molecule is attracted to any other negative charge. In a sample of water, there are many other molecules, so they will tend to align themselves so that the positive charge on the hydrogens of one molecule will be as close as possible to the negative charge on

the <u>oxygen</u> of another molecule. This results in <u>hydrogen bonding</u>, which causes water molecules to be very close to one another. In fact, most molecules that are chemically similar to water are <u>gases</u> at room temperature. Water, however, is a <u>liquid</u> at room temperature because of <u>hydrogen bonding</u>.

10. While the solid phase of most substances <u>sinks</u> in the liquid phase of that same substance, solid water (ice) <u>floats</u> in liquid water. This is because the molecules of liquid water are <u>closer together</u> than the molecules in solid water. This convenient fact allows lakes to <u>freeze</u> from the top down. As a result, a reasonably deep body of water will never <u>completely</u> freeze, because the <u>ice</u> at the surface insulates the water below. This allows fish (and other aquatic animals) to <u>survive</u> the winter.

11. Because of hydrogen bonding, individual water molecules are so strongly <u>attracted</u> to one another that they tend to stay together, even when subjected to an outside force. This gives water its <u>cohesion</u>, which, in turn, causes <u>surface tension</u>, the phenomenon that caused the needle to float in Experiment 4.5. This same phenomenon is exploited by water striders, allowing them to <u>walk</u> on water. The <u>cohesion</u> of water is also what makes it possible for water to travel up through the xylem of a tall plant.

12. Although water's <u>cohesion</u> is strong, it can be overcome. In Experiment 4.6, for example, water "beaded up" on the <u>waxed</u> surface of the glass, because water molecules are attracted to <u>each other</u> more <u>strongly</u> than they are to the molecules that make up wax. However, the water did not "bead up" on the unwaxed glass, because water molecules are attracted to <u>the molecules that make up glass</u> more <u>strongly</u> than they are to each other.

13. When water has ions like calcium and magnesium dissolved in it, we call it <u>hard</u> water. It is <u>not</u> the result of treatment done to make the water safe to drink. Instead, it is the result of the <u>source</u> from which the water is taken. In a <u>water softener</u>, calcium and magnesium ions are "exchanged" with either <u>sodium</u> or <u>potassium</u> ions so that the calcium and magnesium ions are not in the water we drink and use. People who are on strict <u>low-sodium</u> diets should either not soften their water or use more expensive, sodium-free water-softener salts, like <u>potassium chloride</u>.

ANSWERS TO THE SUMMARY OF MODULE #5

1. Water is such a large part of the earth that astronomers often call it the <u>blue</u> planet. The sum of all water on a planet is called its <u>hydrosphere</u>. Of all the planets in our solar system, earth is the <u>only one</u> that has a large quantity of water in its liquid form. This is because the earth has <u>just the right mixture of gases</u> in its atmosphere and is <u>just the right distance</u> from the sun.

2. The vast majority of earth's water supply is contained in the <u>oceans</u> as <u>saltwater</u>. The vast majority of earth's freshwater supply is stored in <u>icebergs</u> and <u>glaciers</u>. The largest source of liquid freshwater is <u>groundwater</u>. Aside from the sources just mentioned, the other major sources of water in the hydrosphere are <u>surface water</u> (not oceans), <u>soil moisture</u>, and <u>atmospheric moisture</u>.

3. The process by which water is continuously exchanged between earth's various water sources is called the <u>hydrologic cycle</u>. In this process, water gets into the atmosphere predominantly by <u>evaporation</u> and <u>transpiration</u>. Soil moisture is usually depleted by either <u>transpiration</u> or <u>groundwater flow</u>. Water vapor in the atmosphere can form a cloud through a process called <u>condensation</u>. Once water is in a cloud, it can fall back to earth as <u>precipitation</u>. When this water falls on land and then runs along the surface into a lake, river, or stream, we call it <u>surface runoff</u>.

4. Evaporation and condensation of a mixture to separate out the mixture's individual components is called <u>distillation</u>. This process is why water from the ocean can eventually end up in a <u>freshwater</u> source, like a lake, river, or stream.

5. The average time a given particle will stay in a given system is called its <u>residence time</u>, and in the hydrologic cycle, it varies considerably from source to source. The average time a molecule of water stays in a swiftly flowing river, for example, is <u>shorter</u> than that of a water molecule in a lake. The average time a molecule of water stays in the atmosphere is much <u>shorter</u> than that of a water molecule in the ocean. The <u>Bible</u> was the first work to mention the hydrologic cycle.

6. The chemical name of the salt you put on your food is <u>sodium chloride</u>. Although this is the majority of salt in the ocean, chemists use the tem "salt" more <u>broadly</u>, and as a result, there are other salts in the ocean. A measure of the mass of dissolved salt in a given mass of water is called <u>salinity</u>.

7. Salt is found in the ocean because the only way water can escape the ocean is through <u>evaporation</u>. As Experiment 5.1 shows, when this happens, the <u>salt</u> is left behind. Thus, the ocean's average salinity <u>increases</u> over time. Nevertheless, the salinity of the ocean does vary. Where rivers dump water into the ocean, for example, the salinity is <u>lower</u> than the average salinity. The average salinity of the ocean indicates it is <u>much younger</u> than even a few million years old.

8. Saltwater freezes at a <u>lower</u> temperature than does freshwater. In fact, putting salt on ice will often <u>melt</u> the ice, because the salt molecules <u>attract</u> water molecules so that they move away from the other water molecules. When the temperature gets low enough, however, even saltwater will freeze, but the salt and water <u>separate</u> as the solution freezes, usually forming solid water that surround little pockets of concentrated saltwater called <u>brine</u>.

9. Icebergs are composed of <u>freshwater</u>. They *do not* form as a result of <u>ocean</u> water freezing. In certain polar regions, the water in the ocean does freeze to form <u>sea ice</u>, but that is not an iceberg.

In fact, icebergs come from <u>glaciers</u>, which are the result of snowfall. When a region is cold enough, the <u>snow</u> does not melt away during the summer. When new snow falls, the old snow gets packed down into what is called <u>firn</u>. As the mass of snow accumulates, it begins to slide to lower elevations, forming a <u>glacier</u>.

10. As glaciers move, they might encounter warmer temperatures, where they begin to <u>melt</u>, feeding various <u>freshwater</u> sources of the hydrosphere. Glaciers in the polar regions often do not encounter warmer temperatures, however, and move all the way to the ocean, where they form <u>ice shelves</u>. When the edge of a glacier advances into the ocean, the ice <u>weakens</u> at some points, and large chunks of ice break off the glacier, floating away in the water. This process, called <u>calving</u>, is what makes an <u>iceberg</u>, approximately 90% of which is <u>under water</u>.

11. Soil moisture can flow down through the soil in a process called <u>percolation</u>. If it travels down far enough, it will reach soil that is completely saturated with water. The line between the saturated and unsaturated soil is called the <u>water table</u>. The depth of this line <u>changes</u> over time. For example, when there is a period of very heavy rains, the depth <u>decreases</u>, and when there are periods of little or no rain, the depth <u>increases</u>.

12. If a lake has a high enough salinity to consider it a saltwater lake, there are <u>no</u> rivers taking water away from the lake. As a result, the only way water can leave is through <u>evaporation</u>. The <u>Dead Sea</u> is one such lake, and it has a much higher salinity than that of the ocean.

13. Water in the atmosphere exists as either <u>humidity</u> or <u>clouds</u>. In order for clouds to form, there must be <u>cloud condensation nuclei</u> upon which water can condense. This condensation occurs because as air expands, it gets <u>cooler</u>. The scientific name for this process is <u>adiabatic cooling</u>. Water in clouds can be either <u>liquid</u> or <u>solid</u>, depending on the temperature.

14. Adiabatic cooling should not be confused with the fact that most things <u>expand</u> when they are heated. When you heat something, you are giving it <u>energy</u>. In adiabatic cooling, air is expanding *without* being given <u>energy</u>.

15. A refrigerator uses a substance that is a <u>gas</u> at room temperature. A compressor in the refrigerator compresses the gas, which <u>heats</u> it up and forms a lot of <u>liquid</u>. Once compressed, the gas is released into a <u>low-pressure</u> portion of the system, which allows it to <u>expand</u>. This <u>cools</u> down the contents of the refrigerator. In addition, the gas that had condensed <u>evaporates</u>, which further cools the system. The pipes that carry the expanded gas are on the <u>inside</u> of the refrigerator, and the pipes that carry the compressed gas are on the <u>outside</u> of the refrigerator.

16. <u>Fog</u> is the result of a cloud forming on the ground. Although this used to be called <u>smog</u>, that term is now generally used to refer to a brownish haze that results from pollution. However, that brownish haze is more properly referred to as <u>photochemical smog</u>.

17. One of the real environmental problems that exists today is water <u>pollution</u>, especially what is occurring to the groundwater supply. Since nearly 50% of the United States gets its <u>drinking water</u> from groundwater sources, it has a direct effect on human health.

ANSWERS TO THE SUMMARY OF MODULE #6

1. The earth is typically divided into five regions: the atmosphere, the hydrosphere, the <u>crust</u>, the <u>mantle</u>, and the <u>core</u>. The deepest region is further subdivided into the <u>outer core</u> and <u>inner core</u>. We have learned about the lowest regions with <u>indirect observation</u>, such as observing how sound waves pass through the earth.

2. The earth's crust is its <u>outermost</u> layer of <u>rocks</u>. It is separated from the mantle by the <u>Mohorovicic discontinuity</u>, which is typically called the <u>Moho</u> for short. We have never been able to drill <u>below</u> the crust. The crust also contains <u>soil</u> and small, solid fragments of rock and other materials called <u>sediment</u>. Many of the rocks of the earth's crust are <u>sedimentary rocks</u>, which are formed when chemical reactions cement sediments together. Other rock types found in the crust are <u>igneous</u> (rock that forms from molten rock) and <u>metatmorphic</u> (rock that has been changed as a result of great pressure and temperature).

3. The mantle is <u>under</u> the crust, and it is separated from the outer core by the <u>Gutenberg discontinuity</u>. Its principal ingredient is <u>silica</u>. Deeper portions of the mantle have a <u>higher</u> temperature than shallower portions of the mantle. The crust and the upper layers of the mantle form the <u>lithosphere</u>, and directly below that is the <u>asthenosphere</u>, where the rock is called <u>plastic rock</u> because it behaves like something between a liquid and a solid.

4. When earthquakes occur, they emit vibrations called <u>seismic waves</u>, which travel through the earth, eventually reaching the surface. They can be detected with <u>seismographs</u>, which can be used to tell how the waves traveled through the inner parts of the earth. This allows scientists to develop <u>models</u> of the earth's mantle and core, which allow us to understand their makeup.

5. The core's principal ingredient is <u>iron</u>. In the outer core, the iron is <u>liquid</u>, but in the inner core, it is <u>solid</u>. Nevertheless, the inner core has a <u>higher</u> temperature than the outer core. The reason the inner core is solid is because of <u>pressure freezing</u>. The boundary between the inner and outer cores is the <u>Lehmann discontinuity</u>.

6. Electrical currents in earth's core are responsible for earth's <u>magnetic field</u>, the strength of which has been <u>decreasing</u> for the past 170 years. In addition, its direction has <u>reversed</u> a few times in the past. The data indicate that at least some of these reversals have happened over a <u>short</u> time period.

7. The earth's magnetic field deflects the vast majority of <u>cosmic rays</u> that come from the sun. Without such protection, <u>life</u> would cease to exist as a result of the <u>energy</u> of these particles. If the earth's magnetic field were too small, <u>too few</u> of them would be deflected. If it were too strong, it would cause deadly <u>magnetic storms</u> that would make life impossible. Thus, the earth has a magnetic field that is <u>just the right strength to support life</u>.

8. There are basically two views of how the electrical currents in the core originated, and they are called the <u>dynamo theory</u> and the <u>rapid-decay theory</u>. The <u>dynamo theory</u> assumes that the earth is billions of years old and is <u>not very successful</u> when compared to the data. The <u>rapid-decay theory</u> assumes that they earth is only thousands of years old and is <u>very successful</u> when compared to the data. The fact that most scientists believe in the <u>dynamo theory</u> in spite of the data indicates that there is no such thing as an <u>unbiased</u> scientist. The <u>rapid-decay theory</u> says that all planets initially had a magnetic field, but some planet's fields have decayed away to nothing by now. The <u>dynamo theory</u>

says that once a planet has a magnetic field, its strength might change, but it will never be completely gone.

9. The theory of <u>plate tectonics</u> views the earth's lithosphere as composed of several "<u>plates</u>" that all move about on the plastic rock of the <u>asthenosphere</u>. When they move away from each other, <u>magma</u> leaks up from the mantle, creating new <u>crust</u>. When they collide, one can slide under the other, generally forming a <u>trench</u> with mountains on one side. When this happens, <u>crust</u> is destroyed as it melts into the mantle. When they collide and neither slides under the other, they <u>buckle</u>, forming mountains. When they <u>slide</u> (or shear) against each other, their edges scrape against each other. This motion can result in severe <u>earthquakes</u>.

10. Many of our observations of <u>earthquakes</u>, <u>mountains</u>, and volcanoes seem to support the theory of plate tectonics. There are deep trenches at the bottom of the oceans, the characteristics of which are well described by the theory that the plates in that region of the earth are moving <u>away from each other</u>. In the end, then, most geologists believe that the plate tectonics theory is <u>correct</u>.

11. The fact that the continents appear as if they fit together like a jigsaw puzzle has led some scientists to speculate that years ago, all the continents were connected in a giant supercontinent, which has been called <u>Pangaea</u>. Evidence to support this idea includes the fact that sections of rock from different continents are <u>very similar</u>, and they "<u>match up</u>" when you put the continents together the way they are assumed to have existed in <u>Pangaea</u>. Although most scientists believe that the plates have always moved <u>slowly</u>, a theory called "catastrophic plate tectonics" uses <u>rapid</u> plate movement as a result of a global catastrophe to explain how the supercontinent split in a short amount of time.

12. Vibration of the earth that results either from volcanic activity or rock masses suddenly moving along a fault is called an <u>earthquake</u>. A fault is the <u>boundary</u> between two sections of rock that can <u>move</u> relative to one another. Wherever a fault exists, there is the possibility of an <u>earthquake</u>.

13. The most successful theory regarding fault-related earthquakes is the <u>elastic rebound theory</u>. In this theory, as rock masses on a fault try to move relative to each other, they get <u>caught up</u> on one another. As a result, they <u>bend</u>. Eventually, the rock masses <u>break free</u> of each other, and they "<u>rebound</u>" to their normal shape.

14. The point where an earthquake begins is called the earthquake's <u>focus</u>. The <u>epicenter</u> is the point on the surface of the earth directly above an earthquake's focus. The study of earthquakes is called <u>seismology</u>, and it uses delicate instruments called <u>seismographs</u> that can measure vibrations that are too small for us to notice. This has led to a scale that classifies earthquakes based on their strength, called the <u>Richter scale</u>. This scale runs from 0 to 10, and each step along this scale is an increase of approximately <u>32</u> in the energy of an earthquake. A earthquake that measures 5 on the Richter scale is <u>32</u> times more energetic than one that measures 4 and <u>32,768 (32x32x32)</u> less energetic than one that measures 8.

15. If a fault exists in which one rock mass is moving up and the other is stationary or moving down, the upward-moving mass of rock will form a <u>fault-block mountain</u>. When two moving rock masses push against each other with extreme force, the crust can bend in an up-and-down, rolling pattern, forming <u>folded mountains</u>. A mountain formed by lava leaking up through the crust from the mantle is a <u>volcanic mountain</u>, while one formed by magma that does not leave the mantle is called a <u>domed mountain</u>.

ANSWERS TO THE SUMMARY OF MODULE #7

1. The term "<u>weather</u>" refers to the condition of the earth's atmosphere (mostly the troposphere) at any particular time. <u>Climate</u>, on the other hand, is a steady condition that prevails day in and day out in a particular region of creation.

2. The principal factors affecting the weather are <u>thermal energy</u>, <u>uneven distribution of thermal energy</u>, and <u>water vapor in the atmosphere</u>.

3. Meteorologists separate clouds into four basic groups: <u>cumulus</u> (fluffy piles of clouds), <u>stratus</u> (layers of clouds), <u>cirrus</u> (high altitude, wispy clouds), and <u>lenticular</u> (lens-shaped clouds). You generally find each type of cloud at a <u>characteristic</u> altitude, but a prefix of "<u>alto</u>" is used to indicate that a cloud type is higher than expected. In general, <u>cirrus</u> clouds form at the highest altitudes, while <u>stratus clouds</u> form at the lower altitudes. A prefix of "<u>nimbo</u>" or a suffix of "<u>nimbus</u>" is also added if the cloud is dark. Dark clouds are the ones that typically bring <u>precipitation</u>.

4. Unusually large, upward-moving wind currents can produce huge, towering <u>cumulonimbus clouds</u> that most people call "thunderclouds." Cirrus clouds are composed of <u>ice</u> instead of liquid water. Precipitation-producing stratus clouds are typically called <u>nimbostratus clouds</u>. Clouds that look like part cirrus/part cumulus clouds are called <u>cirrocumulus clouds</u>, while clouds that look a bit like cumulus clouds but are formed where stratus clouds normally formed are called <u>stratocumulus clouds</u>. Finally, some clouds have the feathery appearance of cirrus clouds, but they form flat layers like that of stratus clouds and are called <u>cirrostratus clouds</u>.

5. Light that comes to the earth from the sun is called <u>insolation</u>, which abbreviates "incoming solar radiation." The earth's <u>distance from the sun</u> and its <u>axial tilt</u> affect how much a region of the earth gets. In addition, cloud cover can <u>reduce</u> the amount of incoming solar radiation. The earth orbits the sun in an oval pattern that mathematicians call an <u>ellipse</u>. When the earth is at its aphelion, it is the <u>farthest</u> it will ever be from the sun. When it is at its perihelion, it is <u>closest</u> to the sun.

6. Because of earth's axial tilt, sunlight shines more directly on the <u>Northern Hemisphere</u> when the earth is at aphelion. Thus, it is <u>summer</u> in the Northern Hemisphere and <u>winter</u> in the Southern Hemisphere at that time. At perihelion, sunlight shines more directly on the <u>Southern Hemisphere</u>. At that time, then, it is <u>winter</u> in the Northern Hemisphere and <u>summer</u> in the Southern Hemisphere.

7. At the two <u>equinoxes</u>, the days are <u>12 hours</u> long in both hemispheres. As the earth moves from the autumnal equinox (spring equinox in the Southern hemisphere) to the winter solstice (summer solstice in the Southern hemisphere), the days in the Northern Hemisphere are <u>shorter</u> than 12 hours and are getting <u>shorter</u>. In the Southern Hemisphere, the days are <u>longer</u> than 12 hours and are getting <u>longer</u>. From the winter solstice (summer solstice in the Southern hemisphere) to the spring equinox (autumnal equinox in the Southern hemisphere), the days in the Northern Hemisphere are <u>shorter</u> than 12 hours and are getting <u>longer</u>. In the Southern Hemisphere, the days are <u>longer</u> than 12 hours and are getting <u>shorter</u>.

8. Most likely, Christ was born in <u>April</u>, not December. However, December 25th is celebrated as Christ's birthday because missionaries tried to link it to a pagan holiday that was called the <u>birthday of the sun</u>.

9. Imaginary lines that run north and south across the earth are called <u>lines of longitude</u>, while imaginary lines that run east and west across the earth are called <u>lines of latitude</u>. The latitude is <u>0</u> at the equator and increases the <u>farther</u> you move away from it. The longitude is <u>0</u> at the prime meridian, which runs through <u>Greenwich, England</u>. It increases the <u>farther</u> you move away from the prime meridian.

10. Hot air <u>rises</u>. As this happens it creates a region of <u>low</u> pressure. Cold air <u>sinks</u>. As this happens, it creates a region of <u>high</u> pressure. These effects cause loops of winds to develop as air tries to move from <u>cool</u> regions of the earth (like the poles) to <u>warm</u> regions of earth (like the equator). These winds are then bent by the <u>Coriolis effect</u>, which stems from the fact that different parts of the earth move at different speeds. The result is prevailing winds in the polar regions called <u>polar easterlies</u>, prevailing winds in the mid latitudes called <u>westerlies</u>, and prevailing winds near the equator called <u>trade winds</u>.

11. Because of the Coriolis effect, a missile fired due north from the equator will end up hitting a target <u>northeast</u> of its launch site, while a missile fired due south from near the North Pole will end up hitting a target <u>southwest</u> of its launch site. The Coriolis effect, however, is <u>not strong enough</u> to significantly affect how water drains in a basin.

12. Prevailing wind patterns can be easily disrupted by <u>local winds</u>. Examples of such winds would be a <u>sea breeze</u> near the ocean shore, which tends to blow during the day, and a <u>land breeze</u>, which tends to blow near the ocean shore during the night.

13. An air mass is a large body of air with relatively uniform <u>pressure</u>, <u>temperature</u>, and <u>humidity</u>. The three basic types of air masses are <u>arctic</u>, <u>polar</u>, and <u>tropical</u>. Artic air masses are very cold and dry. <u>Maritime tropical</u> air masses are warm and moist, while <u>maritime polar</u> air masses are cold and moist. <u>Continental tropical</u> air masses are warm and dry, while <u>continental polar</u> air masses are cold (but not as cold as artic air masses) and dry.

14. A weather front is a <u>boundary</u> between two air masses. The four basic types are <u>cold fronts</u>, <u>warm fronts</u>, <u>stationary fronts</u>, and <u>occluded fronts</u>.

15. When a cold front moves in, <u>cumulonimbus</u> clouds are usually formed by the warm air <u>rising</u> in response to the cold air mass. The temperature in the region tends to <u>decrease</u>. Cold fronts generally carry the most <u>severe</u> weather system, including thunderstorms.

16. When a warm front moves in, the warm air tends to <u>rise</u> above the cooler air that was in the region. This usually causes a progression of clouds from cirrus to <u>cirrostratus</u> to stratus to <u>nimbostratus</u>, which generally heralds a <u>slow</u> and <u>steady</u> rain as well as <u>increased</u> temperatures.

17. Occluded fronts occur when a <u>cold front</u> meets up with a slower-moving <u>warm front</u>. They usually result in slow, steady rains followed by <u>thunderstorms</u>.

18. A stationary front generally results in weather that doesn't <u>change</u> much for a long period of time.

ANSWERS TO THE SUMMARY OF MODULE #8

1. Eventually, all the water that evaporates into the atmosphere falls back to earth, mostly in the form of <u>precipitation</u>. However, water can also leave the atmosphere and return to earth as <u>dew</u> or <u>frost</u>.

2. By far, the most common form of precipitation is <u>rain</u>. There are two theories about how rain forms in clouds. The <u>Bergeron</u> process deals with how rain is formed in cold clouds. The ice crystals in these clouds grow <u>larger</u> until they can no longer remain <u>suspended</u> in the air. As they fall, they typically pick up more <u>ice</u>, growing even heavier. Eventually, these ice crystals become so big that they <u>fragment</u>, which results in several ice crystals falling through the cloud. Each of these fragments, until there are billions of <u>ice crystals</u> falling from the cloud. As they descend, they melt and form <u>rain</u>.

3. In warm clouds, meteorologists think that rain forms according to the <u>collision-coalescence process</u>. In this theory, each cloud contains many water droplets. As <u>air currents</u> in the cloud move these droplets around, they <u>collide</u> with other water droplets. Sometimes the droplets stick together, forming a <u>bigger</u> water droplet. Eventually, a water droplet gets big enough to start <u>dropping</u> through the cloud.

4. Drizzle usually forms in <u>stratus</u> clouds. Sleet is different from freezing rain because sleet is frozen <u>before</u> it hits the ground, while freezing rain is not. <u>Hail</u> is formed when an ice crystal or raindrop is blown back into the cloud by an upward gust of wind. If blown high enough, the raindrop will <u>freeze</u>, or the ice crystal will get <u>larger</u>. Depending on the wind conditions, the ice crystal might be blown back up into the clouds <u>several</u> times. Eventually, it gets so big that the upward gusts of wind are <u>no longer</u> strong enough to push it back up into the clouds, and it falls to the earth. Snow starts out as precipitation from a <u>cold</u> cloud. As the ice crystals fall from the clouds, they <u>absorb more water</u>, freezing and growing into bigger ice crystals.

5. A thunderstorm begins with a current of rising air, called an <u>updraft</u>. As the air rises, water condenses onto cloud condensation nuclei, which actually <u>heats</u> the cloud condensation nuclei, making the current of rising air stronger. This is the <u>cumulus</u> stage of the thunderstorm. Eventually, the water droplets and/or ice crystals in the cloud become too <u>large</u>, and it begins to rain. This marks the <u>mature</u> stage of the thunderstorm. As the rain falls, it causes winds that blow downward, which are called <u>downdrafts</u>. These winds eventually overpower the rising currents of air that started the storm, and the entire area is full of only <u>downdrafts</u>. This marks the <u>dissipation</u> stage of the thunderstorm. A single thunderstorm cell typically lasts for less than <u>30</u> minutes, but a thunderstorm might be composed of <u>many</u> cells so that the storm lasts longer.

6. Lightning forms because a charge <u>imbalance</u> in a cloud causes charge to build up on the <u>ground</u>. The positive charges on the ground attract some negative charges from the cloud, forming a <u>stepped leader</u>. The closeness of the negative charges forces the positive charges up, making the <u>return stroke</u>, which is the most powerful part of the lightning strike. The <u>thunder</u> that you hear is the result of air that has been superheated by the return stroke. Although this kind of lightning (called <u>cloud-to-ground</u> lightning) forms lightning bolts, <u>cloud-to-cloud</u> lightning lights up the sky in big sheets.

7. Tornadoes start as the result of updrafts that form <u>thunderstorms</u>. In the first stage of their development, known as the <u>whirl</u> stage, the updraft of air forming a cumulonimbus cloud begins being hit by winds blowing in a different direction at higher altitudes. Combined with the updraft, this causes a funnel of air to form, with air whirling both around and up. This is often called a <u>vortex</u>. The

funnel of air then touches the ground, starting the <u>organizing</u> stage of the tornado. Once the funnel touches the ground, it sucks debris up into the funnel, which darkens the tornado. This marks the <u>mature</u> stage. It is in this stage that the tornado is most destructive. Eventually, the forces that hold the vortex together dissipate, and the tornado gets smaller, entering its <u>shrinking</u> stage. Finally, the tornado weakens to the point that it is no longer visible, and it slowly dies out in the <u>decaying</u> stage. When tornadoes form over the water, the result is a <u>waterspout</u>, which is <u>weaker</u> than a tornado that forms over land. A <u>dust devil</u> is even weaker, forming as a result of temperature differences between the ground and the air above it.

8. Hurricanes are more properly called <u>tropical cyclones</u>, because they always start in the tropics. They begin as a <u>tropical disturbance</u> that is fed by the warm, moist air of the tropical sea. If the rotating winds reach a sustained speed of 23 miles per hour, it is "upgraded" to a <u>tropical depression</u>. If the winds reach a sustained speed of 39 miles per hour, the depression is "upgraded" again to a <u>tropical storm</u>. Finally, if the winds reach 74 miles per hour, it becomes a full-fledged hurricane. There are <u>five</u> categories of hurricanes, which are based on the wind speeds in the storm. The most pronounced feature of a hurricane is its <u>eye</u>, and the clouds spin around the eye <u>counterclockwise</u> in the Northern Hemisphere and <u>clockwise</u> in the Southern Hemisphere. The eye is actually a place of <u>calm</u> in the midst of the storm.

9. <u>Radar</u> (which stands for "radio detection and ranging"), emits <u>radio</u> waves at a rate of several hundred per second. As those waves encounter objects, they <u>bounce</u> off the objects and head back toward the radar unit. The time it takes for the waves to travel to an object and then bounce back indicates the <u>distance</u> to the object. In addition, differences between the outgoing and returning waves provide information that can determine whether a cloud is made up of <u>ice crystals</u> (a cold cloud) or <u>water droplets</u> (a warm cloud). <u>Doppler radar</u> is a well-known tool in both weather and law enforcement. Traffic police use it to determine the <u>speed</u> of automobiles, while meteorologists use it to measure the <u>speed</u> of winds and air masses.

10. Weather <u>satellites</u> take data continuously all over the world and give us an accurate, <u>global</u> picture of the weather fronts and patterns that exist on a day-to-day basis. They also provide strong evidence that global warming is <u>not</u> happening.

11. Weather data is often summarized on a <u>weather</u> map that allows meteorologists to track fronts and atmospheric pressure. The thin black lines on such a map are called <u>isobars</u>, and they represent regions of equal <u>atmospheric</u> pressure. An "H" on such a map indicates an area of <u>high</u> pressure, while an "L" represents <u>low</u> pressure. Isobars represent <u>increasing</u> pressure the farther they are from an "L" and <u>decreasing</u> pressure the farther they are from an "H."

12. If a thick line on a weather map has only triangles on it, it represents a <u>cold</u> front, and the way the triangles point tell you the <u>direction</u> in which the front travels. If it has only ovals on it, the line represents a <u>warm</u> front, and the side the ovals are on tells you the <u>direction</u> in which the front travels. If the line has both ovals and circles on the same side, it represents an <u>occluded</u> front, and once again, the side that the symbols are on tells you the <u>direction</u> in which the front travels. Finally, if the line has ovals on one side and triangles on another, it represents a <u>stationary</u> front.

ANSWERS TO THE SUMMARY OF MODULE #9

1. Every science relies on the science of <u>physics</u>. As a result, we call it the most <u>fundamental</u> of all the sciences. The science of <u>mechanics</u> is the branch of physics that deals with analyzing and understanding objects in motion, the <u>forces</u> that are applied to those objects, and the <u>energy</u> that exists in them.

2. When studying motion, one must define a <u>reference point</u>, which is a point against which position is measured. If an object's position relative to this point is <u>changing</u>, the object is in motion relative to that point. Because motion depends on the reference point, all motion is <u>relative</u>.

3. The units for speed and velocity are composed of a <u>distance</u> unit divided by a <u>time</u> unit. In base metric units, speed is given in <u>meters/sec</u>. While <u>speed</u> simply tells you how quickly an object is moving, <u>velocity</u> tells you how quickly an object is moving *and* the direction in which it moves. Thus, speed is a <u>scalar</u> quantity, while velocity is a <u>vector</u> quantity. Speed can be calculated with the equation:

$$\text{speed} = \frac{\text{distance traveled}}{\text{time traveled}}$$

4. When objects travel in the same direction, their relative speed is the <u>difference</u> between their individual speeds. When they travel in opposite directions, their relative speed is the <u>sum</u> of their individual speeds.

5. The time rate of change of an object's velocity is its <u>acceleration</u>. The units for this quantity are composed of a <u>distance</u> unit divided by a <u>time</u> unit <u>squared</u>. In base metric units, it is given in meters/sec². It is a <u>vector</u> quantity, because it contains directional information. It can be calculated with the equation:

$$\text{acceleration} = \frac{\text{final velocity} - \text{initial velocity}}{\text{time}}$$

6. An object with an unchanging speed can still have acceleration, provided that its <u>direction</u> is changing. In physics, the term "acceleration" can also mean that an object is <u>slowing down</u>, because acceleration is simply the change in velocity, and a decrease in velocity is still a change. If an object is speeding up, its acceleration is in the <u>same direction</u> as its velocity. If it is slowing down, its acceleration is in the <u>opposite direction</u> as compared to its velocity.

7. When an object falls solely under the influence of gravity, we say that it is in <u>free fall</u>. In such a situation, the acceleration is equal to <u>9.8 meters/second²</u> in metric units and <u>32 feet/second²</u> in English units. This acceleration is <u>independent</u> of the characteristics of the object. Thus, in true free fall, a feather and a bowling ball will fall with the <u>same</u> acceleration.

8. When an object is in free fall, the distance it drops can be calculated with the equation:

$$\text{distance} = \frac{1}{2} \times (\text{acceleration}) \times (\text{time})^2$$

9. Although we generally treat objects falling near the surface of the earth as if they were in free fall, air resistance impedes the fall of all objects. Thus, things don't truly free fall unless there is no air. However, for most objects, the effect of air resistance can be ignored. Thus, when most objects fall near the surface of the earth, we can assume they are in free fall.

10. When doing an experiment in which error is a known problem, you can reduce the effects of error by making several measurements and averaging the results.

11. This problem gives us distance and time and asks for speed. Thus, we need to use Equation (9.1). The problem wants the answer in miles per hour, however. We are given the time in minutes, so we must make a conversion first:

$$\frac{45 \text{ minutes}}{1} \times \frac{1 \text{ hour}}{60 \text{ minutes}} = 0.75 \text{ hours}$$

Now we can use our speed equation:

$$\text{speed} = \frac{20 \text{ miles}}{0.75 \text{ hours}} = 26.7 \ \underline{\frac{\text{miles}}{\text{hour}}}$$

12. Vector quantities contain information about direction, scalar quantities do not. The units tell you what the measurement is of:

a. The units (distance divided by time^2) means this is acceleration. Since there is a direction, this is a vector quantity.

b. The units (distance divided by time) means this is either speed or velocity. Since there is no direction, it is a scalar quantity and is a measurement of speed.

c. The units (distance divided by time) means this is either speed or velocity. Since there is no direction, it is a scalar quantity and is a measurement of speed. Even though the information about slowing is in the measurement, that tells us nothing about direction.

d. The units (distance divided by time) means this is either speed or velocity. Since direction is given, it is a vector quantity and is a measurement of velocity.

13. The initial velocity is 0, and the final velocity is 15 meters per second east. The time is 2.1 seconds. This is a straightforward application of Equation (9.2).

$$\text{acceleration} = \frac{\text{final velocity} - \text{initial velocity}}{\text{time}}$$

$$\text{acceleration} = \frac{15 \ \frac{\text{meters}}{\text{second}} - 0 \ \frac{\text{meters}}{\text{second}}}{2.1 \text{ seconds}} = \frac{15 \ \frac{\text{meters}}{\text{second}}}{2.1 \text{ seconds}} = 7.14 \ \frac{\text{meters}}{\text{second}^2}$$

Since the car sped up, the acceleration is in the same direction as the velocity. Thus, the acceleration 7.14 m/sec^2 east.

14. The rock is in free fall, so we can use Equation (9.3). Since the problem wants the answer in meters, we need to use 9.8 meters per second^2 as the acceleration.

$$\text{distance} = \frac{1}{2} \times (\text{acceleration}) \times (\text{time})^2$$

$$\text{distance} = \frac{1}{2} \times (9.8 \, \frac{\text{meters}}{\text{second}^2}) \times (3.8 \, \text{seconds})^2$$

$$\text{distance} = \frac{1}{2} \cdot (9.8 \, \frac{\text{meters}}{\text{second}^2}) \times (14.44 \, \text{second}^2) = \underline{70.8 \, \text{meters}}$$

15. As the picture shows, the car is behind the truck, but they are both traveling in the same direction. Thus, we get their relative speed by subtracting their individual velocities:

$$\text{relative speed} = 42 \text{ miles per hour - 37 miles per hour} = 5 \text{ miles per hour}$$

Since the car is traveling slower than the truck, the truck is pulling away from the car. Thus, the relative velocity is 5 miles per hour away from each other.

16. Since the velocity is not changing, the acceleration is zero. The time was just put in there to fool you. Remember, acceleration is the change in velocity. With no change in velocity, there is no acceleration.

17. The final velocity is zero (the car ended up stopped) and the initial velocity is 55 miles per hour. It takes 3.5 seconds for the car to stop. The time units do not agree. Since the question asks for the answer to be put in miles per hour^2, we need to convert seconds to hours.

$$\frac{3.5 \, \text{seconds}}{1} \times \frac{1 \, \text{hour}}{3600 \, \text{seconds}} = 0.000972 \, \text{hours}$$

Now we can calculate the acceleration:

$$\text{acceleration} = \frac{\text{final velocity} - \text{initial velocity}}{\text{time}}$$

$$\text{acceleration} = \frac{0 \, \frac{\text{miles}}{\text{hour}} - 55 \, \frac{\text{miles}}{\text{hour}}}{0.000972 \, \text{hours}} = \frac{-55 \, \frac{\text{miles}}{\text{hour}}}{0.000972 \, \text{hours}} = -56{,}584 \, \frac{\text{miles}}{\text{hour}^2}$$

The negative sign tells us that the acceleration is in the opposite direction as is the velocity. This makes sense, since the car slowed down. The acceleration, then, is 56,584 miles/hour^2 north.

18. The rock is in free fall, so we can use Equation (9.3). Since the problem wants the answer in feet, we need to use 32 feet per second^2 as the acceleration.

$$\text{distance} = \frac{1}{2} \times (\text{acceleration}) \times (\text{time})^2$$

$$\text{distance} = \frac{1}{2} \times (32 \, \frac{\text{meters}}{\text{second}^2}) \times (2.3 \, \text{seconds})^2$$

$$\text{distance} = \frac{1}{2} \cdot (32 \, \frac{\text{meters}}{\cancel{\text{second}}^2}) \times (5.29 \, \cancel{\text{second}^2}) = \underline{84.64 \text{ feet}}$$

ANSWERS TO THE SUMMARY OF MODULE #10

1. Sir Isaac Newton discovered <u>three</u> laws of motion, developed a theory describing <u>gravity</u>, did the famous prism experiment that showed white light is composed of many <u>colors,</u> and in order to help his scientific investigations, he developed a new kind of mathematics that we now call "<u>calculus.</u>" He was also a devoutly <u>religious</u> man who spent as much time studying the <u>Bible</u> as he did studying science.

2. Newton's three laws of motion are:

> I. <u>An object in motion (or at rest) will tend to stay in motion (or at rest) until it is acted upon by an outside force.</u>

> II. <u>When an object is acted on by one or more outside forces, the total force is equal to the mass of the object times the resulting acceleration.</u>

> III. <u>For every action, there is an equal and opposite reaction.</u>

3. The tendency of an object to resist changes in its velocity is referred to as <u>inertia</u>. When a bomb is dropped from an airplane, the bomb <u>does not</u> hit the ground directly below where the airplane dropped it. Instead, it continues to move in the <u>direction</u> that the plane was moving when the bomb was dropped, because of Newton's <u>First</u> Law of Motion. Thus, a bomber must drop the bomb <u>before</u> it is above the target.

4. The reason Aristotle made so many mistakes when describing motion is that he did not know about the existence of <u>friction,</u> a force that opposes motion and results from the contact of two <u>surfaces</u>. This force exists because on the atomic scale, all surfaces are <u>rough</u>. This affects how close the molecules can get to one another, which affects how much they are <u>attracted</u> to each other. The more they are <u>attracted</u> to one another, the stronger the frictional force. When this force opposes motion once the motion has already started, we call it <u>kinetic friction</u>. When it opposes the initiation of motion, we call it <u>static friction</u>. Between these two forces, <u>static friction</u> is greater than <u>kinetic friction</u>.

5. A force is essentially a push or a pull exerted on an object in an effort to change that object's <u>velocity</u>. You can calculate force with the equation:

$$\underline{\text{total force} = (\text{mass}) \cdot (\text{acceleration})}$$

The units on force are composed of a <u>mass</u> unit times a <u>distance</u> unit divided by a <u>time</u> unit squared. The standard unit for force is $\dfrac{\text{kg} \cdot \text{m}}{\text{sec}^2}$, which is also called the "Newton."

6. When multiple forces act on an object, forces in the same direction are <u>added</u>, and forces in opposite directions are <u>subtracted</u>. Since friction always opposes motion, the frictional force will always be <u>subtracted</u> from the force that is being used to cause motion.

7. When you fire a gun, it "kicks" back towards you. That "kick" is the result of Newton's <u>Third</u> Law of Motion. When you pull the trigger, you cause a <u>chemical reaction</u> to take place in the

chamber. That reaction produces a lot of <u>heat</u> and <u>gas</u>. The gas is under pressure, so it exerts a force on the <u>bullet</u>, pushing the bullet out at an amazing speed. In response, the <u>bullet</u> pushes back against the gas in the gun.

8. The *total* force comes from Newton's Second Law:

$$\text{total force} = (\text{mass}) \cdot (\text{acceleration})$$

$$\text{total force} = (25 \text{ kg}) \cdot (34.5 \frac{\text{meters}}{\text{second}^2})$$

$$\text{total force} = 862.5 \text{ Newtons}$$

Since we are ignoring friction, this is the *only* force, so it is also our answer. Thus, the force is <u>862.5 Newtons west</u>.

9. Since the velocity is constant, acceleration is zero. After all, acceleration tells us how velocity changes. With no change in velocity, acceleration = 0.

$$\text{total force} = (\text{mass}) \cdot (\text{acceleration})$$

$$\text{total force} = (125.0 \text{ kg}) \cdot (0 \frac{\text{meters}}{\text{second}^2})$$

$$\text{total force} = \underline{0 \text{ Newtons}}$$

Since we are ignoring friction, this is the *only* force, so it is also our answer.

10. Since velocity is constant, that means acceleration is zero. Based on Newton's Second Law (as above), the *total* force on the object is also zero. In this case, however, there are *two* forces acting on the object, and they are opposed. There is the force being applied to the object, and there is the frictional force. Since they oppose one another, you must subtract them. The total force is zero, so:

$$\text{applied force - frictional force} = 0$$

Since the object is moving, we use the kinetic frictional force, which is given as 15 Newtons. Thus:

$$\text{applied force - 15 Newtons} = 0$$

The applied force, then, must be 15 Newtons. Since the object is moving north, that is also the direction of the applied force. As a result, the applied force is <u>15 Newtons north</u>.

11. The <u>static frictional force is 40 Newtons</u>, as that's what it takes to get things moving. Once it is moving, however, it has a 30 Newtons force applied to it and accelerates at 0.1 m/sec². Well, with the mass and the acceleration, we can get the *total* force:

$$\text{total force} = (\text{mass}) \cdot (\text{acceleration})$$

$$\text{total force} = (65 \text{ kg}) \cdot (0.1 \frac{\text{meters}}{\text{second}^2})$$

$$\text{total force} = 6.5 \text{ Newtons}$$

There are two forces at play here. We have the applied force (30 Newtons) and the kinetic frictional force. They oppose one another, so we subtract them to get the total force:

$$30 \text{ Newtons - kinetic frictional force} = 6.5 \text{ Newtons}$$

The kinetic frictional force must be 23.5 Newtons. Since it opposes motion, it is opposite the direction of motion. Thus, the kinetic frictional force is 23.5 Newtons east.

12. To get the object moving, just over 45 Newtons must be exerted, as that will overcome the static frictional force. Once moving, kinetic frictional force is at play. The mass and acceleration give us the *total* force:

$$\text{total force} = (\text{mass}) \cdot (\text{acceleration})$$

$$\text{total force} = (95 \text{ kg}) \cdot (0.5 \frac{\text{meters}}{\text{second}^2})$$

$$\text{total force} = 47.5 \text{ Newtons}$$

That force is the result of the difference between the exerted force and the kinetic frictional force:

$$\text{exerted force - 22 Newtons} = 47.5 \text{ Newtons}$$

Thus, the exerted force must be 69.5 Newtons. Since the motion is to the south, the answer is 69.5 Newtons south.

13. There are a total of 4 forces at play. They combine to give the total force. Let's calculate the total first:

$$\text{total force} = (\text{mass}) \cdot (\text{acceleration})$$

$$\text{total force} = (50 \text{ kg}) \cdot (0.1 \frac{\text{meters}}{\text{second}^2})$$

$$\text{total force} = 5 \text{ Newtons}$$

We are given three of the four forces working on the object. Notice that two of them work together in the direction of the acceleration (35 Newtons east and 45 Newtons east). The other one, however, opposes them. In addition, friction will oppose them. Thus, friction and the 10 Newton west force act together. Therefore we subtract both 10 Newtons and friction from the other two forces to get the total force:

$$35 \text{ Newtons} + 45 \text{ Newtons} - 10 \text{ Newtons} - \text{kinetic friction} = 5 \text{ Newtons}$$

$$70 \text{ Newtons} - \text{kinetic friction} = 5 \text{ Newtons}$$

Kinetic friction must be 65 Newtons. It opposes the motion, so it is <u>65 Newtons west</u>.

14. <u>The static frictional force is 75 Newtons</u>. Once it got moving, the worker keeps it moving at a constant velocity west. This tells us that the acceleration is zero, which means the total force on the rock is zero. Thus, the difference between the exerted force (45 Newtons) and the kinetic frictional force must be zero.

$$45 \text{ Newtons} - \text{kinetic friction} = 0 \text{ Newtons}$$

<u>The kinetic frictional force, then, must be 45 Newtons east</u>, against the motion.

15. The total force on the rock can be calculated from the mass and acceleration:

$$\text{total force} = (\text{mass}) \cdot (\text{acceleration})$$

$$\text{total force} = (800 \text{ kg}) \cdot (0.1 \frac{\text{meters}}{\text{second}^2})$$

$$\text{total force} = 80 \text{ Newtons}$$

This force comes from the two men and friction. The two men are pushing in the same direction, so their forces add and the kinetic frictional force subtracts:

$$200 \text{ Newtons} + 150 \text{ Newtons} - \text{kinetic friction} = 80 \text{ Newtons}$$

$$350 \text{ Newtons} - \text{kinetic friction} = 80 \text{ Newtons}$$

The kinetic frictional force, then, must be <u>270 Newtons south</u>.

ANSWERS TO THE SUMMARY OF MODULE #11

1. The weakest of the four fundamental forces in creation is the <u>gravitational</u> force, and it is always attractive. The <u>electromagnetic</u> force exists between charged particles. The <u>weak</u> force governs certain radioactive processes in atoms. Physicists have actually shown that <u>electromagnetic</u> force and the weak force are different facets of the same force. Thus, scientists call this force the <u>electroweak</u> force. The <u>strong</u> force is responsible for holding the center of the atom (called the <u>nucleus</u>) together. Although this force is strong, its range is very, very <u>small</u>.

2. The three general principles contains in Newton's Universal Law of Gravitation are:

    I. <u>All objects with mass are attracted to one another by the gravitational force.</u>

    II. <u>The gravitational force between two masses is directly proportional to the mass of each object.</u>

    III. <u>The gravitational force between two masses is inversely proportional to the square of the distance between those two objects.</u>

3. Although the gravitational force is <u>weak</u>, it can be substantial when at least one of the objects involved has a large <u>mass</u>. In addition, the gravitational forces exerted by two objects on one another are <u>equal</u>. Thus, a ball is attracted to the earth because <u>earth</u> applies a gravitational force on the ball. At the same time, the <u>ball</u> applies an <u>equal</u> but <u>opposite</u> force on the <u>earth</u>.

4. <u>Centripetal</u> force is the force necessary to make an object move in a circle, and it is always directed <u>perpendicular</u> to the velocity of the object, which means it points to the <u>center</u> of the circle. Since the direction of the object moving in a circle is continually <u>changing</u>, it experiences <u>acceleration</u> regardless of whether or not its speed stays constant. Centrifugal force is <u>not</u> a real force. It is simply a consequence of <u>Newton's First Law of Motion</u>.

5. If the centripetal force operating on an object moving in a circle suddenly disappears, the object begins traveling <u>straight</u>, in the direction it was moving the instant the force disappeared. Centripetal force can be summed up with three basic principles:

    I. <u>Circular motion requires centripetal force.</u>

    II. <u>The larger the centripetal force, the faster an object travels in a circle of a given size.</u>

    III. <u>At a given speed, the larger the centripetal force, the smaller the circle.</u>

6. The gravitational force acts to hold the planets and their moons in an orderly arrangement which we call the <u>solar system</u>. The sun's <u>gravity</u> applies a centripetal force to the planets, allowing them to travel around the sun in roughly <u>circular</u> orbits. The closest planet to the sun is <u>Mercury</u>, and continuing out from there, you find <u>Venus</u>, <u>earth</u>, and <u>Mars</u>. Next you find the solar system's highest concentration of <u>asteroids</u>. As a result, this region is often called the <u>asteroid belt</u>. Beyond that you find <u>Jupiter</u>, <u>Saturn</u>, <u>Uranus</u>, and <u>Neptune</u>. Typically, the planets of the solar system are placed into one of two groups: the <u>inner planets</u> (Mercury, Venus, earth, and Mars) and the <u>outer planets</u> (Jupiter, Saturn, Uranus, and Neptune). Of all the planets,

Venus is the hottest because of its <u>atmosphere</u>. <u>Pluto</u> was once called a planet, but it is now called a <u>dwarf planet</u>. It was demoted when a larger <u>Kuiper Belt Object</u> (KBO) named Eris was found.

7. When an object orbits around a planet, we call that object a <u>satellite</u> of the planet. All planets except Mercury and Venus have at least one natural <u>satellite</u>, but most have more than one. In addition, Saturn, Uranus, Jupiter, and Neptune all have <u>planetary rings</u>, the most pronounced of which are around Saturn. They are actually composed of <u>small bodies</u> of rock, ice, and frozen gases.

8. Variations in a body's motion are called <u>perturbations</u>, and a careful study of them led to the discovery of the planet <u>Neptune</u>. When they happen to an asteroid, it can be thrown out of its standard orbit and towards earth. When it intersects earth's orbit, it is called a <u>meteoroid</u>. When it actually hits earth's atmosphere, it becomes white-hot, making brilliant streaks of light in the sky. At that point, scientists call it a <u>meteor</u>. The intense heat usually breaks it up, except for a few small pieces that fall to the ground and are called <u>meteorites</u>.

9. Comets are called "dirty <u>snowballs</u>" by some physicists because they are mostly composed of dust grains, chunks of dirt, and <u>ice</u>. When a comet passes close to the sun, the solid part of the comet is called the <u>nucleus</u>, and the "fuzzy" atmosphere around the comet is called the <u>coma</u>, which can form a long, glowing tail in the night sky. <u>Short-period</u> comets typically don't go farther from the sun than the planet <u>Jupiter</u> and take less than 200 years to make an orbit. <u>Long-period</u> comets typically have orbits that extend to the planet <u>Neptune</u> or beyond and take more than 200 years to orbit the sun. The <u>Kuiper belt</u> contains many bodies that have some characteristics of comets and is thought by many to be a source of <u>short-period</u> comets. There are, however, problems with that view. Scientists forced to believe that the solar system is billions of years old must also believe in the <u>Oort cloud</u> as a source for <u>long-period</u> comets, although there is no evidence for its existence.

10. There are essentially two theories on what causes the gravitational force: the <u>General Theory of Relativity</u> and the <u>graviton theory</u>. The <u>General Theory of Relativity</u> says that gravity is a consequence of how mass bends both space and time, while the <u>graviton theory</u> states that gravity is a result of the fact that objects with mass exchange particles called <u>gravitons</u>. Most physicists would say that the <u>General Theory of Relativity</u> is better, since it has some direct evidence supporting it.

11. The Greeks thought that all planets (and the sun) orbited around the <u>earth</u>. This is called the <u>geocentric</u> view of the solar system. As time went on, observations just couldn't be made consistent with this view, so scientists (like Copernicus) suggested the <u>heliocentric</u> view that the plants orbit the sun.

12. The only difference is that the distance between the objects was multiplied by 2. The gravitational force decreases when the distance between the objects increases. It is decreased according to the square of that increase. Thus, since the distance is *multiplied by* 4, the force is *divided by* $2^2 = 4$. <u>The new gravitational force, then, is 4 times smaller than the old one.</u>

13. The gravitational force is proportional to mass. If you look at how the masses changed, $mass_1$ was multiplied by 2 (2 x 5 = 10). $Mass_2$, on the other hand, was multiplied by 3 (2 x 3 = 6). Whatever the mass gets multiplied by also multiplies the force. Thus, the force was multiplied by 2 when $mass_1$ was

changed and again by 3 when $mass_2$ was changed. Thus, the new force is $2 \times 3 = $ <u>6 times stronger</u> than the original force.

14. The gravitational force is proportional to mass. Thus, if you multiply the mass, you multiply force by the same amount. In this case, $mass_1$ was multiplied by 2, while $mass_2$ was multiplied by 8. So the force also gets multiplied by 2 and then by 8, for a total of 16. However, the distance also changed. It was multiplied by 4. The gravitational force decreases when the distance between the objects increases. It is decreased according to the square of that increase. Thus, since the distance was *multiplied by* 4, the force is *divided by* $4^2 = 16$. The new force, then, also gets divided by 16. The effect of the mass change was to multiply the force by 16, but the effect of the distance change was to divide the force by 16. Thus, <u>the new gravitational force is the same as the old one</u>.

ANSWERS TO THE SUMMARY OF MODULE #12

1. Because of the genius of <u>James Clerk Maxwell</u>, we now know that the force between charged particles and the force between magnets are both facets of the same force, called the <u>electromagnetic force</u>. As was the case with Newton, Maxwell studied science as a means of serving <u>Christ</u>.

2. The three principles of the electromagnetic force are:

    I. <u>All electrical charges attract or repel one another: like charges repel, while opposite charges attract</u>.

    II. <u>The force between charged objects is directly proportional to the amount of electrical charge on each object</u>.

    III. <u>The force between charged objects is inversely proportional to the square of the distance between the two objects</u>.

3. The electromagnetic force is significantly <u>stronger</u> than the gravitational force. It is produced by the <u>exchange</u> of small "packages" of light called <u>photons</u>. The more charge a particle has, the more <u>photons</u> it can exchange. This tells you why the electromagnetic force between charged particles is directly proportional to the charge of the particle. When you randomly throw a ball at a person, the chance of you hitting that person is <u>inversely</u> proportional to the <u>square</u> of the distance between you. Thus, the ability for <u>charged particles</u> to exchange photons also is inversely proportional to the square of the distance between them.

4. In an atom, there are as many <u>protons</u> (positive charges) as there are <u>electrons</u> (negative charges). As a result, atoms are electrically <u>neutral</u>. When an atom loses <u>electrons</u>, it ends up with a net positive charge and is called a positive <u>ion</u>. When an atom picks up extra <u>electrons</u>, it ends up with a net negative charge and is called a negative <u>ion</u>.

5. When you charge an object by allowing it to come into contact with an object that already has an electric charge, you are charging by <u>conduction</u>. This gives the newly charged object the <u>same</u> type of charge (positive or negative) as the original object. When you charge an object without direct contact between the object and a charge, you are charging by <u>induction</u>, and the newly charged object typically has a charge <u>opposite</u> that of the original charge.

6. A battery <u>stores</u> electrical charge. One side of the battery is a source of electrons, so it is considered <u>negative</u>. The other is the place where the electrons want to go, so it is considered <u>positive</u>. When the two sides of a battery are hooked together with a metal, <u>electrons</u> will flow through the metal from the <u>negative</u> side of the battery to the <u>positive</u> side. A battery's <u>voltage</u> tells you how hard the battery "pushes" <u>electrons</u> from one side to the other.

7. The amount of charge that travels past a fixed point in an electric circuit each second is called the <u>current</u> in the circuit. It is usually measured in <u>Amperes</u>, which are abbreviated as "<u>amps</u>" or "<u>A</u>." Both the <u>current</u> (amps) and the <u>voltage</u> (volts) of a circuit are needed to know how powerful the circuit is.

8. Current that flows from the positive side of the battery to the negative side is called <u>conventional current</u>. This is the way current is drawn in circuit diagrams, even though it is <u>wrong</u>. The ability of a material to impede the flow of charge is called that material's <u>resistance</u>, and it converts the energy of the charge flowing through the circuit into <u>heat</u> and sometimes <u>light</u>. The <u>type</u> of metal, the <u>width</u> of the metal, and the <u>length</u> of the metal all affect its resistance. The longer the metal, the <u>higher</u> the resistance, and the wider the metal, the <u>lower</u> the resistance.

9. The same number of <u>electrons</u> flow out of a toaster as the number that flow into it. The toaster does use up something, however. It uses <u>energy</u>. As electrons flow through a circuit, the collisions they experience convert the <u>energy</u> produced by the electromagnetic force into <u>heat</u>.

10. A circuit that does not have a complete connection between the two sides of the power source is called an <u>open circuit</u>. Current <u>does not</u> flow through an open circuit. This how a switch works. When the switch is "off," the circuit is <u>open</u>, and no current can flow. When the switch is "on," a <u>connection</u> is made, and current begins to flow.

11. When light bulbs are hooked in a circuit in <u>series</u>, one broken light bulb will cause them all to go out. When light bulbs are hooked in <u>parallel</u>, the other light bulbs will still work even if one or more break.

12. All magnetic force results from the movement of <u>charged</u> particles. The atoms of most materials are not <u>aligned</u>, so the electrons in the material have random motion. This causes the individual magnetic fields that result from that motion to <u>cancel each other out</u>. The result, then, is <u>no</u> magnetic behavior. However, certain materials under certain conditions can have their atoms arranged so that the electrons have the <u>same</u> general motion. When that happens, the result is a <u>magnet</u>.

13. All magnets have two poles: the <u>north pole</u> and the <u>south pole</u>. Opposite poles <u>attract</u> one another, and like poles <u>repel</u> one another. Because all magnets have two poles, they are sometimes called <u>dipoles</u>. The strength of a magnet depends on what <u>percentage</u> of the atoms in the material are <u>aligned</u>. The larger the percentage, the <u>stronger</u> the pole of a magnet.

14. The force is proportional to the charge. Thus, if the charge doubles, the force doubles. In this case, both charges doubled, so the force doubles and then doubles again, for a total of being multiplied by 4. However, the distance also changed. It was multiplied by 4. The force decreases when the distance between the objects increases. It is decreased according to the square of that increase. Thus, since the distance was *multiplied by multiplied by* 4, the force is *divided by* $4^2 = 16$. The effect of the charge change was to multiply the force by 4, but the effect of the distance change was to divide the force by 16. Thus, <u>the new force is 4/16, or one-fourth as strong as (4 times smaller than) the old one</u>.

15. The force is proportional to charge. Thus, if the charges are halved, the force is halved. Thus, the force gets divided by 2 and then divided by 2 again for a total of being divided by 4. However, the distance also changed. It was divided by 2. The force increases when the distance between the objects decreases. It is increased according to the square of that decrease. Thus, since the distance was *divided by* 2, the force is *multiplied by* $2^2 = 4$. The new force, then, gets multiplied by 4. The change in charge multiplied the force by 4, but the change in distance divided it by 4. Thus, <u>the change in charge is offset by the change in distance, and the new force is the same as the old one</u>.

16. Conventional current flows from positive to negative:

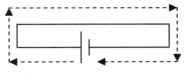

Actual electrons flow from negative to positive:

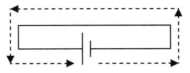

17. When a circuit is open, no current can flow. Thus, there is <u>no current in (a)</u>. However, if the switch is in parallel, current can still flow through the part of the circuit that has a complete connection between the two sides of the battery. Thus, there is current in (b) and (c):

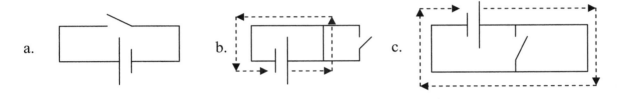

ANSWERS TO THE SUMMARY OF MODULE #13

1. Atoms are made up of <u>protons</u>, <u>neutrons</u>, and <u>electrons</u>. The <u>electron</u> is the smallest and least massive of the three. It also has a <u>negative</u> electrical charge. The <u>proton</u> is next in terms of mass. It is about 2,000 times more massive than the <u>electron</u> and has a <u>positive</u> electrical charge. The <u>neutron</u> is the heaviest of the three, being just a bit more massive than the <u>proton</u>. It has <u>no</u> electrical charge. By itself, the <u>neutron</u> is not stable. If it is not in the nucleus of an atom, it will decay into a <u>proton</u>, an <u>electron</u>, and an antineutrino in a matter of minutes.

2. A <u>model</u> is a schematic description of a system that accounts for its known properties. The <u>Bohr</u> model of the atom has the <u>protons</u> and <u>neutrons</u> packed together in the center of the atom, which is called the <u>nucleus</u>. The <u>electrons</u> orbit the <u>nucleus</u>, much like the planets in the solar system orbit the sun. Although this model is partially <u>wrong</u>, it is still the first model students learn when it comes to the atom. The more correct model is called the <u>quantum-mechanical</u> model, but it is a bit too complex to learn right away. Regardless of the model, we know that atoms (and therefore all of matter) are mostly empty <u>space</u>.

3. One of the most important characteristics of an atom is its number of protons, which is also called its <u>atomic number</u>. This tells you what kind of atom it is. Atoms have equal numbers of protons and <u>electrons</u>, so the atomic number also tells you how many <u>electrons</u> an atom has. The <u>mass number</u> is the sum of the numbers of neutrons and protons in the nucleus, so once you know the number of protons and the <u>mass number</u>, you can figure out how many <u>neutrons</u> are in the nucleus.

4. A collection of atoms that all have the same <u>number</u> of protons is called an "element." On the periodic chart, the chemical symbol for an element is usually the first <u>one</u> or <u>two</u> letters from the <u>English</u> or <u>Latin</u> name of the element. Looking at the periodic chart, you can see that oxygen (O) has an atomic number of <u>8</u>. This means it has <u>8</u> electrons and <u>8</u> protons. An ^{18}O atom, then, has <u>10</u> neutrons, while ^{16}O has <u>8</u> neutrons. Since ^{16}O and ^{18}O all have the same number of protons but different numbers of neutrons, they are <u>isotopes</u>. In the same way, of the following list of atoms: ^{40}Ar, ^{40}Ca, ^{41}K, ^{41}Ca, ^{45}Sc, and ^{42}Ca, the isotopes are <u>^{40}Ca, ^{41}Ca, and ^{42}Ca</u>.

5. If you were to draw an atom of ^{31}P according to the Bohr model, you would start by drawing a nucleus that had <u>15 protons</u> and <u>16 neutrons</u>. It would also have <u>2</u> electrons orbiting the nucleus in the nearest orbit, <u>8</u> electrons orbiting the nucleus in the next orbit out, and <u>5</u> electrons orbiting the nucleus in a third orbit that was farther from the nucleus. In the same way, the Bohr model of ^{84}Sr would have <u>38 protons</u> and <u>46 neutrons</u> in the nucleus. There would be <u>2</u> electrons in the first Bohr orbit, <u>8</u> electrons in the next, <u>18</u> electrons in the third Bohr orbit, and <u>10</u> electrons in the fourth Bohr orbit.

6. Although protons are positively charged and thus should <u>repel</u> one another, experiments have shown that they exist packed together inside the <u>nucleus</u>. This led scientists to speculate that there is a <u>nuclear</u> force that is attractive at very short distances and is strong enough to overcome the repulsive electromagnetic force between protons. Hideki Yukawa showed that the exchange of tiny particles called <u>pions</u> could account for such a force, and he gave a rough prediction of their mass. The detection of <u>short</u>-lived particles with just that mass confirmed the existence of the force.

7. The nuclear force is a short-range force because pions exist for only a <u>short</u> time. Thus, if two protons (or a proton and a neutron) want to exchange a pion, they must do it <u>quickly</u>. The nuclear force is actually

a manifestation of the <u>strong</u> force, which is also manifested in the exchange of <u>gluons</u> between quarks. This allows <u>protons</u> and <u>neutrons</u> to exist.

8. The weak force governs <u>radioactivity</u>. An atom with a nucleus that is not stable is called a <u>radioactive</u> isotope. The three main ways unstable nuclei can decay is through <u>beta</u> decay (where a neutron turns into a proton, <u>electron</u>, and antineutrino), <u>alpha</u> decay (where the nucleus emits a ^{4}He nucleus), and <u>gamma</u> decay (where energy is released in the form of a high-energy photon).

9. The hydrogen isotope known as "tritium" (^{3}H) undergoes beta decay. The daughter product is <u>^{3}He</u>. When the isotope ^{32}P undergoes beta decay, the daughter product is ^{32}S. When ^{133}Xe undergoes beta decay, the daughter product is <u>^{133}Cs</u>. In each case, a <u>beta particle</u> (electron) and an antineutrino are also produced.

10. When ^{238}U undergoes alpha decay, the daughter product is <u>^{234}Th</u>. When ^{222}Rn undergoes alpha decay, ^{218}Po is produced. When ^{210}Po undergoes alpha decay, the daughter product is <u>^{206}Pb</u>. In each case, a <u>^{4}He</u> nucleus is also produced.

11. When ^{60}Ni is produced by the beta decay of <u>^{60}Co</u>, it has excess energy. The ^{60}Ni gets rid of that excess energy by radioactive decay, but it stays ^{60}Ni. Thus, it decays by <u>gamma</u> decay.

12. The <u>half-life</u> of a radioactive isotope is the time it takes for half of the original sample to decay. Consider, for example, ^{239}Np, which has a half-life of 2 days. If you start with 1,000 grams of ^{239}Np, you will have <u>500</u> grams left after 2 days and <u>250</u> grams left after 4 days. In 10 days, you would have <u>31.25</u> grams left. In <u>20</u> days, you would have only about 0.977 grams left.

13. Even though a sample of radioactive isotope never really goes away completely, at some point, the amount of radioactive isotope left is so small that it can be <u>ignored</u>. If we keep a radioactive sample around long enough, then, it will <u>cease</u> to be radioactive, for all practical purposes.

14. Radioactive dating is the process by which scientists use the <u>radioactive decay</u> of certain substances to determine how old an object is. For example, in carbon dating, scientists use the fact that ^{14}C decays with a half-life of 5,700 years. Because living organisms continually exchange ^{14}C with their surroundings, while an organism is alive, it contains the same amount of ^{14}C as does the <u>atmosphere</u> around it. When the organism dies, however, that exchange <u>ceases</u>, and the amount of ^{14}C begins to decrease as a result of radioactive decay. Thus, if you know how much ^{14}C was in the <u>atmosphere</u> when an organism died, you can determine how long ago death occurred by looking at the difference between the ^{14}C in the dead organism and the amount that was in the atmosphere when it died. The difference is assumed to be the result of <u>radioactive decay</u>, and using the known half-life, you can determine the <u>time</u> that elapsed since the organism died.

15. The main problem with carbon dating is determining the <u>amount</u> of ^{14}C in the organism when it died. Scientists can use <u>tree rings</u> to measure the amount of ^{14}C in the atmosphere at a given year, and that gives them the ability to make a good assumption about the <u>amount</u> of ^{14}C in the organism when it died. However, the oldest tree ring analyzed in this way is <u>3,000</u> years old, so carbon dating is really only reliable for things that are <u>3,000</u> years old or younger.

16. Other radioactive dating methods use similar <u>assumptions</u>, and the fact that many radioactive dates are in conflict with each other or with generally accepted dates indicates the <u>assumptions</u> are poor.

ANSWERS TO THE SUMMARY OF MODULE #14

1. In a wave, there are both <u>crests</u> (the highest points on the wave) and <u>troughs</u> (the lowest points on the wave). The distance between the crests (or the distance between the troughs) is called the <u>wavelength</u>, and it is symbolized with the Greek letter λ. The height of the wave is the called the <u>amplitude</u>. The <u>frequency</u> of a wave indicates how many waves hit a certain point every second.

2. Frequency and wavelength can be related to one another through the equation:

$$f = \frac{v}{\lambda}$$

In this equation, "f" is the <u>frequency</u> of the wave, "v" is the <u>speed</u> of the wave, and λ is the <u>wavelength</u>. The units for <u>frequency</u> are 1/sec, which are typically abbreviated as <u>Hz</u>.

3. There are two basic forms that waves can take. A <u>transverse wave</u> is a wave with a direction of propagation that is perpendicular to its direction of oscillation. A <u>longitudinal wave</u> is a wave with a direction of propagation that is parallel to its direction of oscillation. In a <u>longitudinal wave</u>, the places where the medium "bunches up" are called <u>compressions</u>, while the "spread out" sections are called <u>rarefactions</u>.

4. <u>Sound waves</u> are longitudinal waves that generally oscillate <u>air</u>. When those waves reach the <u>tympanic</u> membrane of the ear, the membrane vibrates. Those vibrations are then transmitted to your <u>brain</u>, and your brain interprets them as <u>sound</u>.

5. The speed of sound in air is dependent on the air's <u>temperature</u>, and it can be found using the equation:

$$v = (331.5 + 0.6 \cdot T)\frac{m}{sec}$$

Since the speed of light is significantly <u>faster</u> than the speed of sound, you can see a faraway event <u>before</u> you hear any sound associated with it. Sound travels <u>faster</u> in liquids than it does in gases, and it travels <u>faster</u> in solids than it does in liquids.

6. If an object travels in a medium faster than the speed of sound in that medium, we say that the object is traveling at a <u>supersonic</u> speed. Typically, we use the <u>Mach</u> number to denote such speeds. A rocket traveling at <u>Mach</u> 3, for example, is traveling at three times the speed of sound. The sound produced as a result of an object traveling faster than sound is called a <u>sonic boom</u>.

7. The <u>pitch</u> of a sound wave is governed primarily by its frequency, while the volume is determined mostly by its <u>amplitude</u>. When a singer sings low notes, for example, the sounds waves she produces have <u>low</u> frequency. When she sings high notes, the sound waves she makes have a <u>high</u> frequency. When the singer is singing softly, the sound waves she produces have a <u>small</u> amplitude, and when she sings loudly, the sound waves she makes have a <u>large</u> amplitude.

8. Longitudinal waves with frequencies that can be detected by the human ear are called <u>sonic</u> waves. Waves with frequencies higher than what the human ear can sense are called <u>ultrasonic</u> waves, and waves with frequencies lower than what the human ear can detect are called <u>infrasonic</u> waves.

9. The fact that the pitch of a car's horn changes as the car passes by you is a result of the <u>Doppler Effect</u>. This effect exists because as a sound source moves, the waves in front of the source <u>bunch</u> together, producing a wave with a <u>higher</u> frequency than what you would hear if the source were stationary. The waves behind a moving source are <u>spread</u> out, resulting in a frequency <u>lower</u> than what you would hear if the source were stationary.

10. The bel scale measures the <u>intensity</u> of a sound wave, which is determined by the <u>amplitude</u>. In this scale, each unit corresponds to a factor of <u>10</u> increase in the intensity of the sound wave. Thus, a sound wave that measures 7 bels is <u>1,000</u> times more intense than a sound wave that measures 4 bels. The more common measurement associated with this scale is the decibel. It takes <u>10</u> decibels to make a bel. As a result, a sound measuring 80 decibels has an intensity of <u>8</u> bels.

11. Sound waves used to probe the inside of the earth are typically <u>infrasonic</u> waves, while sound waves used to measure distances and image things inside the human body are typically <u>ultrasonic</u> waves. Another use of such waves is <u>sonar</u>, a technique used both by the military and by animals such as bats. Despite our best efforts, however, the bat's <u>sonar</u> is significantly more <u>sophisticated</u> than anything made as a result of human science and technology.

12. We use the frequency equation to solve this:

$$f = \frac{v}{\lambda}$$

$$f = \frac{5 \, \frac{\cancel{m}}{sec}}{1.5 \, \cancel{m}} = 3.333 \, \frac{1}{sec}$$

The frequency is <u>3.333 Hz</u>.

13. We use the frequency equation to solve this:

$$f = \frac{v}{\lambda}$$

$$f = \frac{150 \, \frac{\cancel{m}}{sec}}{0.15 \, \cancel{m}} = 1,000 \, \frac{1}{sec}$$

The frequency is <u>1,000 Hz</u>.

14. This is an application of Equation (14.2)

$$v = (331.5 + 0.6 \cdot T) \, \frac{m}{sec}$$

$$v = (331.5 + 0.6 \cdot 22) \, \frac{m}{sec}$$

$$v = (331.5 + 13.2) \, \frac{m}{sec} = \underline{344.7 \, \frac{m}{sec}}$$

15. Because light travels so quickly, we can assume that the lightning was formed at the instant in which you see it. The time delay, then, is simply the time it took for the sound to travel from the point at which the lightning was created to you. First, we need to know the speed of sound. This can be determined by Equation (14.2):

$$v = (331.5 + 0.6 \cdot T) \, \frac{m}{sec}$$

$$v = (331.5 + 0.6 \cdot 11) \, \frac{m}{sec}$$

$$v = (331.5 + 6.6) \, \frac{m}{sec} = 338.1 \, \frac{m}{sec}$$

Now that we have the speed and we know the time, we can calculate distance:

$$\text{distance traveled} = (\text{speed}) \times (\text{time traveled})$$

$$\text{distance traveled} = (338.1 \, \frac{m}{sec}) \times (2.3 \, sec) = \underline{777.63 \, m}$$

16. We can solve this one the same as the problem above.

$$v = (331.5 + 0.6 \cdot T) \, \frac{m}{sec}$$

$$v = (331.5 + 0.6 \cdot 18) \, \frac{m}{sec}$$

$$v = (331.5 + 10.8) \, \frac{m}{sec} = 342.3 \, \frac{m}{sec}$$

Now that we have the speed and we know the time, we can calculate distance:

$$\text{distance traveled} = (\text{speed}) \times (\text{time traveled})$$

$$\text{distance traveled} = (342.3 \frac{m}{sec}) \times (1.3 \text{ sec}) = \underline{444.99 \text{ m}}$$

17. To solve problems like this, we must first convert the numbers to bels.

$$\frac{20 \cancel{\text{ decibels}}}{1} \times \frac{1 \text{ bel}}{10 \cancel{\text{ decibels}}} = 2 \text{ bels}$$

$$\frac{60 \cancel{\text{ decibels}}}{1} \times \frac{1 \text{ bel}}{10 \cancel{\text{ decibels}}} = 6 \text{ bels}$$

Each step in the bel scale is a factor of 10. There are 4 bels between 2 and 6, so we must take 10 times itself 4 times: 10x10x10x10 = 10,000. Thus, the second sound is <u>10,000</u> times more intense than the first.

18. Each step in the bel scale is a factor of 10 in intensity. If the amplifier can multiply the intensity by 1000, that is 10x10x10. Thus, three bels are added. The sound going in is:

$$\frac{30 \cancel{\text{ decibels}}}{1} \times \frac{1 \text{ bel}}{10 \cancel{\text{ decibels}}} = 3 \text{ bels}$$

The final sound, then, is 6 bels. We must convert that back to decibels to answer the question, however:

$$\frac{6 \cancel{\text{ bels}}}{1} \times \frac{10 \text{ decibels}}{1 \cancel{\text{ bel}}} = \underline{60 \text{ decibels}}$$

ANSWERS TO THE SUMMARY OF MODULE #15

1. In the <u>particle</u> theory of light, a beam of light behaves the same as a stream of particles that all move in the same direction. In the <u>wave</u> theory, light is considered a wave. Modern scientists believe that light has a <u>dual</u> nature, acting both like a <u>particle</u> and a <u>wave</u>. In the <u>quantum-mechanical</u> theory, light is basically viewed as tiny packets of waves.

2. Because of the work of James Clerk Maxwell, a light wave is considered a <u>transverse</u> wave composed of an oscillating <u>electric</u> field and a <u>magnetic</u> field that oscillates perpendicular to the <u>electric</u> field. As a result, light waves are typically called <u>electromagnetic waves</u>.

3. Although the speed of light does <u>not</u> depend on temperature, it does depend on <u>the substance</u> through which the light passes. In liquids light travels <u>slower</u> than it does in air, and in solids light travels <u>slower</u> than it does in liquids. Einstein's Special Theory of Relativity says that the speed of light in a vacuum represents the <u>maximum</u> speed that can ever be attained by any object that has mass.

4. The wavelength of visible light determines its <u>color</u>. The 7 basic colors in the rainbow, in order of *increasing* wavelength, are: <u>violet</u>, <u>indigo</u>, <u>blue</u>, <u>green</u>, <u>yellow</u>, <u>orange</u>, and <u>red</u>. While the light we can see with our eyes is part of the <u>visible</u> spectrum of light, the collection of all electromagnetic waves in creation is called the <u>electromagnetic</u> spectrum.

5. Ultraviolet light, X-rays, and gamma rays have wavelengths <u>shorter</u> than visible light. Although they have so much energy that they can <u>kill</u> living tissue, there are some uses for them. Infrared light, microwaves, television waves, and radio waves have wavelengths <u>longer</u> than visible light.

6. When light bounces off an obstacle, we call it <u>reflection</u>. When this happens, the angle of incidence will <u>equal</u> the angle of <u>reflection</u>. Images form in a mirror because light that <u>reflects</u> off a mirror is detected by an eye, and the <u>brain</u> that receives the eye's electrical impulses extends the light <u>backwards</u> to form an image behind the mirror. The image is, of course, <u>fake</u>. It is simply a result of the fact that the brain interprets light as traveling in a <u>straight</u> line.

7. When a wave enters an obstacle, it usually <u>bends</u> in response to its change in speed. When this happens, we say that the wave has been <u>refracted</u>. When light enters a substance in which it must slow down, the light ray will bend <u>toward</u> a line perpendicular to the surface it strikes. When light enters a substance in which it speeds up, the light ray will bend <u>away from</u> a line perpendicular to the surface it strikes. When you are looking at an object underwater, it will appear to be in a location that is <u>different</u> from its actual location, because the light rays <u>refract</u> when they leave the water to hit your eyes.

8. When white light hits a water droplet in the air, some <u>reflects</u> and some <u>refracts</u> into the water droplet. Since the amount of <u>refraction</u> depends partially on the <u>wavelength</u> of the light involved, this separates the white light into its colors. As the light travels through the water droplet, it eventually hits the other side. A portion of the light <u>refracts</u> out of the water droplet, but a portion <u>reflects</u>. The <u>reflected</u> light travels to the other side of the droplet, where a portion is <u>reflected</u> and a portion is <u>refracted</u>. The portion that is <u>refracted</u> has its wavelengths separated even more, because the amount of <u>refraction</u> depends in part on the wavelength of light. With this second <u>refraction</u>, the light has been separated enough for us to distinguish the colors. As a result, the best way to see a rainbow is for the sun to be <u>behind</u> you.

9. When a lens focuses horizontally traveling light rays through a single point (called the <u>focal</u> point), we call it a <u>converging</u> lens. The sides of such a lens have a <u>convex</u> shape. When a lens bends horizontally traveling light rays so that they begin traveling away from each other, we call it a <u>diverging</u> lens. The sides of such a lens have a <u>concave</u> shape.

10. The most elegant application of a converging lens in all of God's creation can be seen in the <u>eye</u>. The eye is covered by a thin, transparent substance called the <u>cornea</u>. Light enters the eye through the <u>pupil</u>, which is essentially an opening left by the <u>iris</u>. When you are in the presence of <u>bright</u> light, the <u>iris</u> closes down to allow only a small amount of light into the eye. When there is <u>little</u> light, the <u>iris</u> opens wide, allowing a larger percentage of the light in. Once light enters the <u>pupil</u>, it is focused by a <u>converging</u> lens. The light is focused onto the <u>retina</u>, which is made up of light-sensitive cells called <u>rods</u> and <u>cones</u>. When these cells sense light hitting them, they send electrical messages down the <u>optic</u> nerve to the <u>brain</u>, which decodes the messages and forms them into images.

11. In order to focus light onto the retina, the eye's lens actually <u>changes shape</u>. This is done through the action of the <u>ciliary muscle</u>, which squeezes or expands the lens. Human science cannot make a lens as <u>sophisticated</u> as that which you find in the eye.

12. If you are <u>nearsighted</u>, your eye can use its ciliary muscle to change the lens enough to keep the image of objects close to you focused on the retina. However, as the object moves <u>farther</u> away, the lens's <u>focal</u> point cannot be changed enough to keep the image there. As a result, the image gets blurry because the light is focused <u>in front</u> of the retina. Because light is being refracted too strongly, a <u>diverging</u> lens can be used to correct this problem. When you are <u>farsighted</u>, your eye's lens can adjust to objects far away, but it cannot focus on objects that are close. This is because the eye refracts light too <u>weakly</u>. A <u>converging</u> lens must be used to correct this problem.

13. The cones in your retina are used to detect the <u>color</u> of the light you are seeing. Some cone cells are sensitive only to <u>low</u>-frequency visible light (red light), while others are sensitive to <u>medium</u>-frequency visible light (green light), while still others are sensitive to <u>high</u>-frequency visible light (blue light).

14. The additive primary colors are <u>red</u>, <u>blue</u>, and <u>green</u>. Television screens and computer monitors <u>add</u> these colors to make all the colors you see. Red and green, for example, add in equal parts to make <u>yellow</u>, while <u>blue</u> and red add in equal parts to make magenta.

15. While the additive primary colors can add to make all the colors you see, the <u>subtractive</u> primary colors are used for inks and paints. These three colors are <u>cyan</u>, <u>magenta</u>, and <u>yellow</u>. When equal amounts of yellow and magenta inks are mixed, for example, the result is <u>red</u> ink. When equal amounts of cyan and <u>yellow</u> inks are mixed, the result is <u>green</u> ink.

16. The colors we see from objects are a result of the wavelength of light that <u>reflects</u> off them and hits our eyes. The dye that colors a shirt, for example, uses the <u>subtractive</u> primary colors to determine what wavelengths <u>reflect</u> off the shirt and hit our eyes. As a result, if a blue shirt is put in a dark room and magenta light is shined on it, the shirt will appear to be <u>magenta</u>, since the blue dye reflects both cyan and magenta light. If a yellow light were shined on the blue shirt in a dark room, it would appear <u>black</u>.

ANSWERS TO THE SUMMARY OF MODULE #16

1. Although the sun is a <u>main</u> sequence star, there are several things that make it very <u>special</u> when compared to other stars in the universe. The vast majority of stars in the universe exist in <u>multiple-star</u> systems, where stars orbit each other. This would cause severe <u>temperature</u> changes in any planet that orbited such a system, making it extremely difficult for <u>life</u> to exist on any such planet. There are much bigger stars in the universe, but if the sun were as big as they are, it would <u>engulf</u> the earth in its orbit! In addition, there are stars that are smaller than the sun. However, if the earth were to orbit such a star, it would have to be <u>very close</u> in order to get enough energy to support life. If a planet got that close to such a star, the large <u>gravitational</u> forces it would experience would make it far too dangerous to support life. Of the stars in this general region of the universe, the sun is in the top 10% in terms of its <u>mass</u>. The combination of the sun's mass and size, then, make it the <u>perfect</u> star to support life on earth.

2. The sun is essentially a big ball of <u>hydrogen</u> and <u>helium</u> gas. The part of the sun that we can see is called the <u>photosphere</u>. Underneath that you find the <u>convection zone</u>, and underneath that, the <u>radiative zone</u>. The deepest part of the sun, however, is its <u>core</u>, where nuclear <u>fusion</u> reactions turn <u>hydrogen</u> into <u>helium</u>. Those reactions produce <u>energy</u>, which is what makes the heat and light that the sun emits.

3. Every now and again, sudden and intense variations in the brightness of the <u>photosphere</u> occur. These variations are called <u>solar flares</u>, and they send enormous amounts of energy to the earth in a short amount of time, which can <u>disrupt</u> satellites, radio communications, and even power grids. Although the surface of the sun's photosphere is a place of violent activity, it is "<u>tame</u>" compared to other, similar stars in the universe. Single stars with roughly the same size and composition of the sun release solar flares that are 100 to 100 million times more <u>powerful</u> than even the most violent solar flares that we have seen coming from the sun.

4. When a large nucleus is split into smaller nuclei, it is called <u>nuclear fission</u>. This process can result in a large amount of <u>energy</u>, and it is the basis of how nuclear <u>power</u> plants make electricity. When two or more small nuclei fuse to make a bigger nucleus, it is called <u>nuclear fusion</u>, and that is what powers the sun.

5. Nuclear fission reactions require a <u>neutron</u> and a large nucleus, and they produce two or more smaller nuclei and several <u>neutrons</u>. Because of this situation, one nuclear reaction can start <u>several more</u> nuclear reactions. If there is a <u>critical</u> mass of the large nucleus, this can lead to a <u>chain reaction</u>. If there is a lot more than a <u>critical</u> mass of the large nucleus, the <u>chain reaction</u> can get out of control, resulting in a nuclear explosion.

6. While nuclear power is reasonably cheap and will last a long, long time, it can be <u>dangerous</u>. A nuclear power plant <u>cannot</u> explode, because there isn't enough of the large nucleus to allow the chain reaction to get that out-of-control. However, if the control systems fail, a <u>meltdown</u> can occur, which is what happened to the <u>Chernobyl</u> nuclear power plant in the Soviet Union in 1986. Nuclear fission also produces <u>radioactive</u> byproducts, and there is no clean way to dispose of them. Although nuclear fission can be dangerous and polluting, it is not clear that it is any more dangerous and polluting than other forms of energy production. Coal-burning power plants, for example, dump pollution into the <u>air</u>, and coal mining has resulted in more than 100,000 <u>deaths</u> in the U.S. since 1900.

7. The <u>spectral letter</u> of a star is determined by the star's temperature. The <u>brightness</u> of a star as it appears in the night sky is called the star's apparent magnitude, while its <u>brightness</u> after being corrected for the <u>distance</u> from the earth to the star is its absolute magnitude. Plotting the absolute magnitude of stars versus their temperature makes a <u>Hertzsprung-Russell Diagram</u>, which is used by astronomers to classify stars.

8. On the Hertzsprung-Russell Diagram, the <u>main sequence</u> stars form a roughly diagonal band that goes from the upper left of the graph to the lower right of the graph. <u>Supergiant</u> stars form a diffuse band at low absolute magnitudes and various temperatures. <u>Red giant</u> stars are found at low temperatures and absolute magnitudes of about zero to -5. <u>White dwarf</u> stars are found at high temperatures and high absolute magnitudes.

9. Main sequence stars are the most <u>common</u> in the universe. The more massive a main sequence star is, the <u>lower</u> its absolute magnitude. White dwarves seem to be the next most <u>common</u>. They are <u>small</u> and not very <u>bright</u>. They are, however, very <u>massive</u>. Supergiants are the <u>largest</u> stars in the universe. They are very bright and seem to be <u>rare</u> in the universe. Red giants are <u>huge</u> stars that are very bright. They produce their energy by nuclear <u>fusion</u>, but it is different from the kind that takes place in main sequence stars.

10. Stars with absolute magnitudes that change are called <u>variable stars</u>, and there are two main kinds: <u>pulsating variables</u> and <u>novas</u>. <u>Novas</u> are exploding stars, the most extreme being <u>supernovas</u>, which expand rapidly and brighten enormously and then fade away <u>permanently</u>. The debris left over from such an explosion is a cloud of bright gases called a <u>nebula</u>. A <u>pulsating variable</u>, on the other hand, regularly expands and contracts without losing <u>mass</u>. This changes its <u>brightness</u> on a regular basis. One particular type of pulsating variable, the <u>Cepheid variable</u>, is important, as it is used to measure universal distances that are too long to be measured with the more precise method known as <u>parallax</u>. <u>Eclipsing</u> binary stars vary in apparent brightness not because their absolute magnitude changes but because one of the two stars can <u>block</u> our view of the other.

11. A light year is defined as the <u>distance light could travel along a straight line in one year</u>. Since light travels quickly, this is a very <u>long</u> distance.

12. A large ensemble of stars, all interacting through the gravitational force and orbiting around a common center is called a <u>galaxy</u>. There are four main types: <u>spiral galaxies</u>, <u>elliptical galaxies</u>, <u>lenticular galaxies</u>, and <u>irregular galaxies</u>.

13. Our galaxy, the <u>Milky Way</u>, is a <u>spiral galaxy</u>. Our sun is on the inner edge of the <u>Orion</u> arm of that galaxy. Our galaxy is part of a small group of about 30 galaxies, which is known as the <u>Local Group</u>. This group of galaxies is on the outer edge of a cluster of galaxies known as the <u>Virgo Cluster</u>.

14. The fact that light coming from distant galaxies has longer wavelengths than expected is referred to the <u>red</u> shift. Although it was once thought that this was a <u>Doppler</u> shift, it is now considered evidence that the universe is <u>expanding</u>. Most astronomers think the universe is <u>expanding</u> without a <u>center</u>. However, some think that there is a <u>center</u> which is roughly marked by earth's solar system. If this is true, light from galaxies that are billions of light years away could have traveled to earth while just a few <u>thousand</u> years passed on earth.

TEST FOR MODULE #4

1. Define the following terms:

a. Electrolysis
b. Polar molecule
c. Solvent
d. Solute
e. Cohesion
f. Hard water

2. What is the chemical formula for water?

3. Some metals tend to absorb oxygen but not hydrogen. Suppose such a metal was covering the battery in an electrolysis experiment like Experiment 4.1. Which is the more likely erroneous result the experiment would yield for the chemical formula of water: HO_2 or H_4O?

4. Why is water a liquid at room temperature when all other chemically similar substances are gases at room temperature?

5. Carbon dioxide (CO_2) is one of the gases that we exhale when we breathe. Carbon monoxide (CO) is a poisonous gas associated with burning things under conditions of low oxygen. How many atoms are in one molecule of CO_2? How many atoms are in one molecule of CO?

6. An important component of gasoline is octane, which is composed of molecules that have eight carbon atoms (C) and eighteen hydrogen atoms (H). What is the chemical formula of octane?

7. Why are water molecules polar?

8. If a substance does not dissolve in water, is it most likely ionic, polar, or nonpolar?

9. If a substance dissolves in water, will it dissolve in vegetable oil, a nonpolar substance?

10. Is hard water the result of a city's water treatment process?

TEST FOR MODULE #5

1. Define the following terms:

a. Transpiration
b. Condensation
c. Residence time
d. Percolation
e. Adiabatic cooling

2. Where does the majority of earth's water reside?

3. What is the largest source of liquid freshwater on the planet?

4. What water source is a molecule of water in once it has gone through transpiration?

5. Water was in the ocean and is now in a cloud. What two hydrologic cycle processes happened in order to make that transfer?

6. Where is the residence time longer: in the ocean or in a fast-moving stream?

7. If a lake has no means of getting rid of water except evaporation, does it contain saltwater or freshwater?

8. What do the oceans tell us about the age of the earth?

9. An enormous amount of ocean water in the polar region freezes. Does it form an iceberg? Why or why not?

10. What process in the hydrologic cycle is responsible for making glaciers?

11. What causes the temperature change that allows for condensation, which makes most clouds?

12. If a sample of gas is compressed and nothing else is allowed to change, what will happen to the temperature of the gas?

13. If there is a lot more rain than normal in an area over an extensive length of time, what happens to the depth of the water table?

14. Why is groundwater pollution so hard to trace back to its original source?

TEST FOR MODULE #6

1. Define the following terms:

a. Sedimentary rock
b. Plastic rock
c. Fault
d. Epicenter

2. Label the sections (a-e) of the earth shown in the figure:

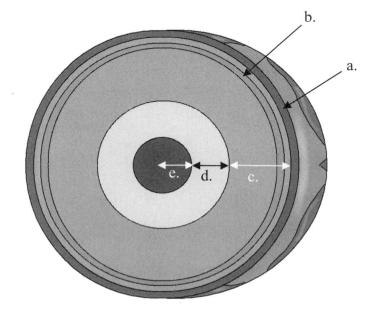

3. What have scientists observed in order to learn about earth's interior?

4. Between what two regions of the earth can you find the Moho?

5. What causes the earth's magnetic field?

6. What two theories attempt to explain the earth's magnetic field? Which theory is the most scientifically valid?

7. What major benefit do we derive from the earth's magnetic field?

8. In a survey of the deep ocean, sonar measurements detect a deep trench on the bottom that runs as far as the instruments detect. What is the most likely cause of the trench?

9. The earthquake activity of two regions on earth is measured. The first region sits near the middle of one of the plates in the earth's crust, while the other is very near a boundary between two plates. Which will (most likely) have the greatest earthquake activity?

10. Many powerful earthquakes are followed later by less-powerful earthquakes called "aftershocks." If an earthquake measures 6 on the Richter scale and is followed by an aftershock that measures 4, how many times more energy was released in the original earthquake as compared to the aftershock?

11. If a region of the earth has a lot of volcanic activity, what kinds of mountains do you expect to find there?

12. Many scientists think that at one time, all the continents might have fit together to form a supercontinent. What is the name of this supercontinent?

TEST FOR MODULE #7

1. Define the following terms:

    a. Aphelion
    b. Perihelion
    c. Coriolis effect
    d. Air mass
    e. Weather front

2. Identify the clouds in the following pictures: *Photos from www.clipart.com*

a.   b.   c.

3. Of the 3 main factors that influence weather, which is mostly responsible for winds?

4. What are dark cumulus clouds called?

5. If an area receives a large amount of insolation, is it likely to be warm or cold?

6. In the Northern Hemisphere, are the day lengths greater than or less than 12 hours between the winter solstice and the spring equinox? Are the day lengths increasing or decreasing during that time?

7. Why is the Northern Hemisphere in winter when the earth is closest to the sun?

8. Suppose you are at the equator and want to fire a missile at a target due north of your location. Would you aim the missile north, northwest, or northeast in order to ensure it hits the intended target?

9. Without two specific factors, the global wind patterns would be simple. They would blow from the poles to the equator. What two factors shape the global winds into the complex patterns that we actually see?

10. What causes the wind in a certain region to be different from what we expect based on the global patterns we see?

11. An air mass is dry and warm. What kind of air mass is it?

12. You watch the sky as cirrus clouds form followed by stratus and nimbostratus clouds. Do you expect a violent rainstorm or a long, lighter rain?

TEST FOR MODULE #8

1. Define the following terms:

a. Updraft
b. Insulator

2. The same cloud precipitates snow on a mountain and rain in the nearby valley. Does the Bergeron process or the collision-coalescence theory best describe the process causing precipitation from that cloud?

3. What is the dew point? What two factors influence it?

4. A thunderstorm cell is raining, and there is no updraft. In what stage is the thunderstorm cell? Will there be hail at this point in the thunderstorm?

5. If the mature stage of a thunderstorm lasts for 30 minutes maximum, why can thunderstorms rain heavy sheets of rain for several hours?

6. Why don't you see lightning from nimbostratus clouds?

7. What happens first in a lightning bolt: a return stroke or a stepped leader?

8. How does lightning cause thunder?

9. Is it possible for sheet lightning to strike a person?

10. A tornado is in its organization stage. Has it touched the ground yet?

11. What differentiates a tropical storm from a tropical disturbance?

12. Where is the calmest part of a hurricane?

(TEST CONTINUES ON THE NEXT PAGE)

Given the following weather map, answer questions 13 - 16.

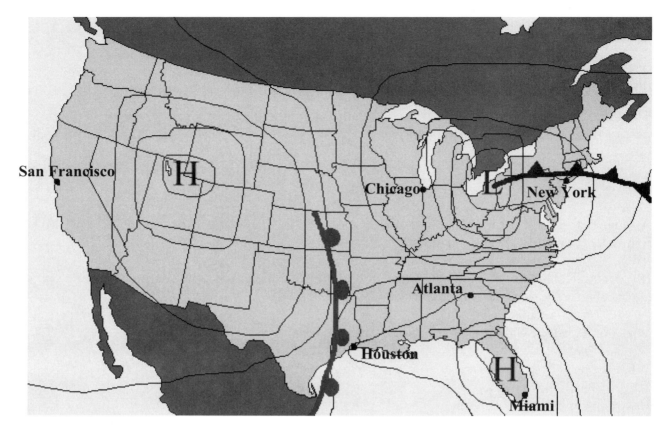

13. Is the atmospheric pressure in Houston, TX higher, lower, or equivalent to that in Atlanta, GA?

14. Is the atmospheric pressure in Chicago, IL higher, lower, or equivalent to that in New York, NY?

15. What city listed on the map might have been experiencing thundershowers at the time this map was drawn?

16. What city listed on the map should expect warmer weather?

TEST FOR MODULE #9
(There are 60 seconds in a minute and 3600 seconds in an hour.)

1. Define the following terms:

    a. Reference point
    b. Vector quantity
    c. Scalar quantity
    d. Acceleration
    e. Free fall

2. Why must one use a reference point to determine whether or not an object is in motion?

3. After a visit to your grandmother's house, you get in your car to go home. You are in the front passenger's seat and your mother is driving. As you back out of your grandmother's driveway, she stands outside, waving good-bye.

    a. Who is in motion relative to you?
    b. Who is motionless relative to you?

4. How many miles per hour does a car travel if it makes a 40-mile trip in 30 minutes?

5. What is the velocity of a bicycle (in meters per second) if it travels 1 kilometer west in 4.1 minutes?

6. You are looking in a scientist's lab notebook and find the following unlabeled measurements. In each case, determine what physical quantity the scientist was measuring.

    a. 12.1 meters per second
    b. 31.2 feet
    c. 14 millimeters per hour to the west
    d. 4.5 yards per minute^2 north

7. An eagle swoops down to catch a baby rabbit. Luckily for the rabbit, he sees the eagle and runs. An all-out chase ensues with the rabbit running east at 5.4 meters per second and the eagle pursuing at 4.4 meters per second. What is the relative velocity of predator and prey?

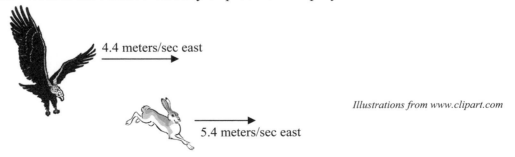

4.4 meters/sec east

Illustrations from www.clipart.com

5.4 meters/sec east

8. What is the acceleration of an object that moves with a constant velocity?

9. A skier reaches the bottom of a slope with a velocity of 12 meters per second north. If the skier comes to a complete stop in 3 seconds, what was her acceleration?

10. A car goes from 0 to 60 miles per hour north in 5 seconds. What is the car's acceleration?

11. A person standing on a bridge over a river holds a rock and a ball in each hand. He throws the ball down towards the river as hard as he can and at the same time simply drops the rock. After both have left the person's hand, does one have a greater acceleration? If so, which one?

12. Why does a dropped feather hit the ground later than a rock dropped at the same time?

13. A physics student climbs a tree. To measure how high she has climbed, she drops a rock and times its fall. It takes 1.3 seconds for the rock to hit the ground. How many feet has she climbed?

TEST FOR MODULE #10

1. Define the following terms:

a. Inertia
b. Friction
c. Kinetic friction
d. Static friction

2. State Newton's three laws of motion.

3. A pilot is flying a mission to drop bombs on an enemy airfield. The plane is flying high and fast to the north, and the city is due north. Should the pilot drop the bombs before the plane is over the airfield, when the plane is over the airfield, or after the plane has passed the airfield?

4. A cruel boy has placed a mouse on the outer edge of a disk. He slowly starts to spin the disk, accelerating it faster and faster until the disk and mouse are both spinning around at an alarming rate. What will happen to the mouse if the boy suddenly stops the disk without touching the mouse: will the mouse continue to spin like it was before; will the mouse stop with the disk; or will the mouse start moving straight, skidding off the disk?

5. An ice cube (mass = 1.0 kg) slides down an inclined serving tray with an acceleration of 4.0 meters per second^2. Ignoring friction, how much force is pulling the ice cube down the serving tray?

6. A baseball player (mass = 75 kilograms) is running north towards a base. In order to avoid being tagged by the ball, the baseball player slides into the base. If his acceleration in the slide is 5.0 meters per second^2 south, what is the kinetic frictional force between the baseball player and the ground?

7. A man pushes a heavy cart. If the man exerts a force of 200 Newtons on the cart to keep it moving at a constant velocity, what is the frictional force between the cart and the ground? Is this kinetic friction or static friction?

8. You are looking through a physicist's laboratory notebook and notice two numbers for the friction between a block of wood and a laboratory bench. The numbers are 8 Newtons and 11 Newtons. Which refers to static friction and which refers to kinetic friction?

9. A woman pushes a box (mass = 30 kilograms). The static friction between the box and the ground is 20 Newtons, while the kinetic friction is 7 Newtons. How much force must the woman exert to get the box moving? With what force must she push the box in order to get it to accelerate at 1.0 meters per second^2 to the west?

10. When baseball players hit the ball hard enough, their bats can sometimes break. What is exerting a force on the bat, causing it to break?

TEST FOR MODULE #11

1. A child is twirling a toy airplane on a string at a constant speed:

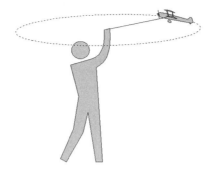

Draw the velocity of the plane and the force that it experiences.

2. Which is the weakest of the fundamental forces?

3. A student drops a ball, and it begins to fall due to the force of gravity that the earth exerts on it. What is the equal and opposite force demanded by Newton's Third Law of Motion?

4. The gravitational force between two objects ($mass_1$ = 5 kg, $mass_2$ = 2 kg) is measured when the objects are 10 centimeters apart. If the 5-kg mass is replaced with a 20-kg mass and the 2-kg mass is replaced with a 12-kg mass, how does the new gravitational attraction compare to the first one that was measured?

5. The gravitational force between two objects ($mass_1$ = 10 kg, $mass_2$ = 6 kg) is measured when the objects are 12 centimeters apart. If the distance between them is increased to 36 centimeters, how does the new gravitational attraction compare to the first one that was measured?

6. Two moons orbit two different planets, but they orbit their planets at the same distance. If the first one takes 3 months to make an orbit and the second takes 1 year, which is being subjected to the greatest gravitational attraction?

7. List the outer planets.

8. Which planet in the solar system receives the least amount of insolation?

9. Which of the following orbits is more likely that of a comet that is relatively easy to see in the night sky?

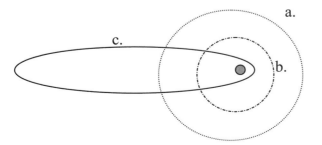

10. What part or parts of a comet come and go during a comet's orbit?

11. What is the Kuiper belt considered to be a source of?

12. What theory says that gravity is caused by the exchange of particles between objects with mass?

13. Suppose a scientist determines that there are only two fundamental forces in nature: the electroweak force and the strong force. Which of the two current theories of gravity does this mean is true?

TEST FOR MODULE #13
(You may use the periodic chart to answer these questions.)

1. Define the following terms:

a. Nucleus
b. Atomic number
c. Mass number
d. Isotopes
e. Element
f. Radioactive isotope

2. Which is larger, an electron or a proton? Which is negatively charged?

3. What causes the nuclear force, and why can this force act only over very short distances?

4. List the number of protons, neutrons, and electrons in the following atoms:

a. ^{48}Ca   b. ^{124}Sn   c. ^{109}Ag

5. Draw an illustration of what the Bohr model says a ^{23}Na atom looks like.

6. Which of the following atoms are isotopes?

$$^{144}Ce, \, ^{144}Nd, \, ^{144}Sm, \, ^{145}Nd$$

7. A radioactive isotope has a half-life of 3 hours. If a scientist has 30 grams of the isotope, how much is left after 15 hours?

8. What is the daughter product in the beta decay of ^{144}Ce?

9. What is the daughter product in the alpha decay of ^{220}Rn?

10. What is the daughter product in the gamma decay of ^{239}U?

11. If a piece of paper is placed between a radioactive isotope and a person, which kind of radioactive particle will the person be protected from?

TEST FOR MODULE #14

1. Define the following terms:

a. Transverse wave
b. Longitudinal wave
c. Supersonic speed
d. Sonic boom
e. Pitch

2. A recorder is a woodwind instrument in which the player blows into a tube, setting up a wave in the tube. A musician has 2 recorders. The first one is rather short, and the second one is significantly longer. Which recorder is capable of playing notes with the lowest pitch?

3. A popular science fiction movie was advertised with the slogan, "In space, no one can hear you scream." Why is this a true statement?

4. Do sound waves oscillate parallel to or perpendicular to the direction in which the wave travels?

5. What is the speed of sound in air that has a temperature of 25 °C?

6. A sound wave traveling through 17 °C air has a wavelength of 2 meters. What is the frequency of the sound wave?

7. Which waves have the longest wavelength: sonic waves, infrasonic waves, or ultrasonic waves?

8. During a thunderstorm, the temperature is 10 °C. If you see a lightning strike and then hear the thunder 2 seconds later, how far away did the lightning strike?

9. A ship blows its horn. Some of the sound waves travel through the air and then hit the water. Will the sound waves travel faster in the water or in the air?

10. A man and woman are singing a duet. The man sings the low notes, and the woman sings the high notes. The woman, since she is singing the melody, is louder than the man. Are the wavelengths of the man's sound waves longer than, shorter than, or the same size as those of the woman? Are the frequencies of the man's sound waves lower or higher than those of the woman? What about the amplitudes of the waves? What about the speed of the sound waves?

11. A jet travels through 10 °C air at Mach 3. What is its speed in meters per second?

12. You are driving in a city that has a siren which sounds for 30 seconds every day at noon. You stop at a stoplight and then hear the sound of the siren. The stoplight then turns green, and you start driving. As you speed up, you notice that the pitch of the siren keeps getting higher. Are you driving toward or away from the siren? (Assume that the true pitch of the siren stays constant.)

13. An amplifier takes a 30 decibel sound and turns it into an 80 decibel sound. How many times larger is the intensity of the sound waves coming out of the amplifier as compared to the intensity of the sound waves going into the amplifier?

TEST FOR MODULE #15

1. Define the following terms:

a. Electromagnetic wave
b. The Law of Reflection

2. Describe light according to the currently accepted theory. Be very detailed.

3. Light is traveling in air and suddenly hits a glass window. Does the light speed up, slow down, or continue traveling at the same speed when it enters the glass.

4. Which has a higher frequency: green light or yellow light?

5. Which has longer wavelengths: ultraviolet rays or radio waves?

6. Draw what happens to the light ray when it hits the right side of the aquarium.

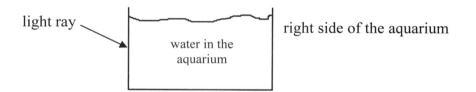

7. You are on a diving board looking down into a pool. You see a quarter at the bottom of the pool, about three feet in front of you. Why is the quarter really not 3 feet in front of you?

8. You want to concentrate light coming from a weak light source by focusing it all on a single point. Would you use a converging or diverging lens to do this?

9. What does the eye do to change its focus?

For problems 10-12, remember that blue + green = cyan, red + blue = magenta, and red + green = yellow.

10. If a computer monitor makes a green dot and then puts a blue dot in the same place, what color will you see?

11. If cyan and magenta ink is mixed, what color ink will you get?

12. If you shine red light on a yellow shirt, what color will the shirt appear? Assume the shirt color is done with dyes that use the subtractive primary colors.

TEST FOR MODULE #16

1. Define the following terms:

    a. Nuclear fusion
    b. Nuclear fission
    c. Critical mass
    d. Absolute magnitude
    e. Apparent magnitude
    f. Light year
    g. Galaxy

2. What nuclear process occurs in the sun's core?

3. A ^{4}He nucleus and a ^{7}Li nucleus collide and form a ^{10}B nucleus and a neutron. Is this nuclear fusion or nuclear fission?

4. A scientist studies a process in which a neutron strikes a ^{216}Pu nucleus to make a ^{104}Cd nucleus, a ^{110}Pd nucleus, and three neutrons. If the scientist measures the mass of the ^{216}Pu and the original neutron and then subtracts the mass of the ^{104}Cd nucleus, the mass of the ^{110}Pd nucleus, and the mass of the three neutrons, will the scientist get a positive number, a negative number, or zero?

5. Is it possible for a nuclear power plant to experience a nuclear explosion? Why or why not?

6. Using the H-R diagram below, classify the following stars:

a. Magnitude 5, Spectral Letter G   b. Magnitude 2, Spectral Letter K
c. Magnitude 11, Spectral Letter B   d. Magnitude -6, Spectral Letter A

Spectral Letter

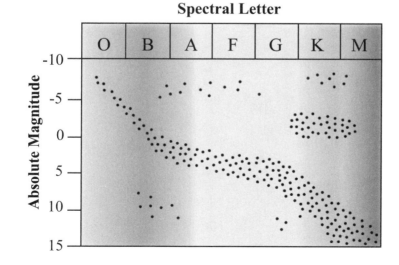

7. Which is the brightest of the stars in problem #6?

8. Which is the hottest of the stars in problem #6?

9. What are the two main types of variable stars?

10. What are Cepheid variables, and why are they important in astronomy?

11. What is the name of the galaxy to which earth's solar system belongs? What type of galaxy is it?

12. What is the red shift?

SOLUTIONS TO THE TEST FOR MODULE #1

1. (3 pts – one for each definition)

a. <u>Atom</u> – The smallest chemical unit of matter

b. <u>Molecule</u> – Two or more atoms linked together to make a substance with unique properties

c. <u>Concentration</u> – The quantity of a substance within a certain volume

2. (1 pt) <u>Sulfur dioxide is composed of molecules.</u> We can tell this because it does not have the properties of sulfur (a yellow powder) but we are told that it contains sulfur atoms. The only way an atom can give up its properties is to become a part of a molecule.

3. (1 pt) <u>The statue will have a copper color underneath the green powder.</u> We learned from Experiment 1.1 and the subsequent discussion that the green powder we see on statues and the like is copper hydroxycarbonate. This molecule needs copper to form. Thus, the statue must have copper on it in order for the copper hydroxycarbonate to have formed on it.

4. (1 pt) <u>Picture A represents a bunch of molecules, whereas picture B is a representation of a bunch of atoms.</u>

5. (1 pt) <u>This is a measurement of mass.</u> The metric unit gram measures mass, regardless of the prefix in front of it.

6. (1 pt) The prefix "<u>kilo</u>" is used to represent the number "1,000."

7. (2 pts – one for the conversion relationship and one for the actual conversion) First, we convert the number to a fractional form:

$$\frac{1.6 \text{ m}}{1}$$

Next, since we want to convert from meters to centimeters, we need to remember that "centi" means "0.01." So one centimeter is the same thing as 0.01 meters. Thus:

$$1 \text{ cm} = 0.01 \text{ m}$$

That's our conversion relationship. Since we want to end up with cm in the end, we must multiply the measurement by a fraction that has meters on the bottom (to cancel the meter unit that is there) and cm on the top (so that cm is the unit we are left with). Remember, the numbers next to the units in the relationship above go with the units. Thus, since "m" goes on the bottom of the fraction, so does "0.01." Since "cm" goes on the top, so does "1."

$$\frac{1.6 \text{ m}}{1} \times \frac{1 \text{ cm}}{0.01 \text{ m}} = 160 \text{ cm}$$

Therefore, 1.6 m is the same as <u>160 cm</u>.

8. (2 pts – one for the conversion relationship and one for the actual conversion) First, we convert the number to a fractional form:

$$\frac{0.12 \text{ kL}}{1}$$

Next, since we want to convert from kiloliters to L, we need to remember that "kilo" means "1,000." So one kiloliter is the same thing as 1,000 liters. Thus:

$$1 \text{ kL} = 1,000 \text{ L}$$

That's our conversion relationship. Since we want to end up with L in the end, we must multiply the measurement by a fraction that has kiloliters on the bottom (to cancel the kL unit that is there) and L on the top (so that L is the unit we are left with):

$$\frac{0.12 \cancel{\text{ kL}}}{1} \times \frac{1,000 \text{ L}}{1 \cancel{\text{ kL}}} = 120 \text{ L}$$

Thus, 0.12 kL is the same as <u>120 L</u>.

9. (2 pts – one for having a fraction with a "1" and a "14.59" in it and one for the actual conversion)

$$\frac{45.1 \cancel{\text{ kg}}}{1} \times \frac{1 \text{ slug}}{14.59 \cancel{\text{ kg}}} = 3.09 \text{ slugs}$$

There are <u>3.09 slugs</u> in 45.1 kg. Note that I rounded the answer. The real answer was "3.091158328," but there are simply too many digits in that number. When you take chemistry, you will learn about significant figures, a concept that tells you where to round numbers off. For right now, don't worry about it. If you rounded at a different spot than I did, that's fine.

10. (1 pt) <u>The second bucket has the more powerful cleaner</u>. Remember, concentration is what is important. Both buckets have the same amount of ammonia in them, but the ammonia in the second bucket is contained in a smaller volume. Thus, the second bucket is more concentrated in ammonia.

Total possible points: 15

SOLUTIONS TO THE TEST FOR MODULE #2

1. (5 pts – one for each definition)

a. Humidity - The moisture content of air

b. Absolute humidity – The mass of water vapor contained in a certain volume of air

c. Relative humidity – The ratio of the mass of water vapor in the air at a given temperature to the maximum mass of water vapor the air could hold at that temperature, expressed as a percentage.

d. Greenhouse effect – The process by which certain gases (principally water vapor, carbon dioxide, and methane) trap heat that radiates from earth

e. Parts per million – The number of molecules (or atoms) of a substance in a mixture for every 1 million molecules (or atoms) in that mixture

2. (1 pt) Water evaporates slowly under conditions of high humidity. Remember, the higher the humidity, the more water vapor that is already in the air. This makes it harder to put more water vapor in the air, which is what evaporation does.

3. (1 pt) No, sweating will not cool you down. When the humidity is 100%, water will not evaporate, and that is what gives sweat its cooling effect.

4. (1 pt) Nitrogen makes up the majority of the air we inhale. See Figure 2.5.

5. (1 pt) Nitrogen makes up the majority of the air we exhale. See Figure 2.5.

6. (1 pt) The substance will burn the fastest in the second trial. In that trial, the oxygen concentration is more than twice that of the first trial. After all, the room's air supply would have been about 21% oxygen. Since the second trial used an air mixture that was 50% oxygen, the larger concentration of air would result in faster combustion.

7. (1 pt) Ozone blocks the ultraviolet light from the sun. Without it, life could not exist on the planet.

8. (1 pt) Ground-level ozone concentrations should be decreased. Remember, ozone is a poison. We do not want to breathe it. We want it all up in the ozone layer.

9. (1 pt) No. The average temperature of the earth has remained rather steady since 1925. See Figure 2.6.

10. (2 pts – one for having a fraction with "10,000" and "1" in it and one for the actual conversion) Remember, we know the relationship between percent and ppm, so we can convert using the factor-label method.

$$\frac{1\%}{1} \times \frac{10,000 \text{ ppm}}{1\%} = 10,000 \text{ ppm}$$

A concentration of 1% is the same as 10,000 ppm.

11. (2 pts – one for having a fraction with "10,000" and "1" in it and one for the actual conversion) Remember, we know the relationship between ppm and percent. We can therefore just use the factor-label method to figure out the answer.

$$\frac{0.018 \cancel{\text{ppm}}}{1} \times \frac{1\%}{10,000 \cancel{\text{ppm}}} = 0.0000018\%$$

A concentration of 0.018 ppm is equal to <u>0.0000018 %</u>.

12. (1 pt) <u>Catalytic converters reduced the concentration of carbon monoxide.</u>

Total possible points: 18

SOLUTIONS TO THE TEST FOR MODULE #3

1. (7 pts – one for each definition)

a. <u>Atmosphere</u> – The mass of air surrounding a planet

b. <u>Barometer</u> - An instrument used to measure atmospheric pressure

c. <u>Homosphere</u> – The lower layer of earth's atmosphere, which exists from ground level to roughly 80 kilometers (50 miles) above sea level

d. <u>Heterosphere</u> – The upper layer of earth's atmosphere, which exists higher than roughly 80 kilometers (50 miles) above sea level

e. <u>Jet streams</u> – Narrow bands of high-speed winds that circle the earth, blowing from west to east

f. <u>Heat</u> – Energy that is transferred as a consequence of temperature differences

g. <u>Temperature</u> – A measure of the energy of random motion in a substance's molecules

2. (1 pt) <u>You would study the troposphere</u> because that's where the majority of weather phenomena are.

3. (1 pt) <u>You would study the stratosphere</u> because that's where the ozone layer is.

4. (1 pt) <u>It must have come from the heterosphere.</u> Air in the homosphere is all 78% nitrogen, 21% oxygen, and 1% other.

5. (3 pts – one for each region) <u>The troposphere, stratosphere, and mesosphere</u> are in the homosphere.

6. (1 pt – ½ for each region) <u>The thermosphere and exosphere are in the heterosphere.</u>

7. (1 pt) <u>The height of the column will decrease.</u> Remember, there is a column of liquid there in the first place because of an imbalance between air pressure inside and outside the column. As air seeps in, that imbalance will decrease, resulting in a smaller column. If the leak is big enough to allow as much air in as possible, the level of liquid inside and outside the column will be the same.

8. (1 pt) Temperature increases with increasing altitude in the <u>stratosphere</u> because of the ozone layer.

9. (1 pt) <u>The "ozone hole" is a seasonal phenomenon located only at the South Pole because ozone cannot be depleted by CFCs without the aid of the Polar Vortex.</u> Since the Polar Vortex is seasonal and only exists at the South Pole, the same can be said for the "ozone hole."

10. (1 pt) <u>Heat is energy that is being transferred. To freeze water, energy must be transferred from the water to the surroundings.</u> Thus, water freezes because of heat!

11. (1 pt) The average speed of the molecules would <u>decrease</u>. Remember, temperature measures the average energy of molecules, which is directly related to their speed. Since temperature decreases with increasing altitude in the troposphere, the average energy of the molecules in the troposphere decreases, which means their speeds decrease as well.

Total possible points: 19

SOLUTIONS TO THE TEST FOR MODULE #4

1. (6 pts – one for each definition)

a. Electrolysis – The use of electricity to break a molecule down into smaller units

b. Polar molecule – A molecule that has slight positive and negative charges due to an imbalance in the way electrons are shared

c. Solvent – A liquid substance capable of dissolving other substances

d. Solute – A substance that is dissolved in a solvent

e. Cohesion – The phenomenon that occurs when individual molecules are so strongly attracted to each other that they tend to stay together, even when exposed to tension

f. Hard water - Water that has certain dissolved ions in it – predominately calcium and magnesium ions

2. (1 pt) H_2O

3. (1 pt) H_4O would be the more likely erroneous result. If the oxygen was being absorbed by the metal covering on the battery, there would be less collected in the tube. This would make it look like water molecules had a lot *less* oxygen. HO_2 would be a chemical formula that indicates *more* oxygen.

4. (1 pt) Hydrogen bonding keeps the water molecules close together.

5. (2 pts – one for each number) There are three atoms in a carbon dioxide molecule and two atoms in a carbon monoxide molecule. Remember, each capital letter signifies an atom, and if there is no subscript, that means there is only one atom.

6. (1 pt) The chemical formula is C_8H_{18}.

7. (1 pt) Water molecules are polar because both the oxygen atoms and the hydrogen atoms are fighting over the electrons they are supposed to be sharing. Oxygen can pull on those electrons harder, so it gets more than its fair share of electrons, making it slightly negative. Since the hydrogen atoms get less than their fair share of electrons, they end up slightly positive.

8. (1 pt) It is most likely nonpolar because water dissolves both polar and ionic substances.

9. (1 pt) It will not dissolve in vegetable oil because if it dissolves in water, it is either ionic or polar. Either way, such a substance will not dissolve in a nonpolar liquid like vegetable oil.

10. (1 pt) No. Hard water is the result of the calcium-containing compounds in the region from which the water is taken.

Total possible points: 16

SOLUTIONS TO THE TEST FOR MODULE #5

1. (5 pts – one for each definition)

a. <u>Transpiration</u> – Evaporation of water from plants

b. <u>Condensation</u> – The process by which a gas turns into a liquid

c. <u>Residence time</u> – The average time a given particle will stay in a given system

d. <u>Percolation</u> – The process by which water moves downward in the soil, toward the water table

e. <u>Adiabatic cooling</u> – The cooling of a gas that happens when the gas expands with no way of getting more energy

2. (1 pt) It resides in the <u>oceans</u>.

3. (1 pt) <u>Groundwater</u> is the largest source of liquid freshwater.

4. (1 pt) It is in the <u>atmosphere</u>. Remember, transpiration takes water from plants and puts it in the atmosphere.

5. (2 pts – one for each answer) <u>Evaporation</u> allowed it to leave the ocean and <u>condensation</u> put it in the cloud.

6. (1 pt) The residence time is longest where the least amount of water exchange takes place. Since evaporation is the only way out of the ocean, it will take a long time for a molecule of water to leave the ocean. Thus, <u>the ocean</u> has a longer residence time.

7. (1 pt) <u>It contains saltwater</u>. If the only means of losing water is evaporation, the salts continue to concentrate, making saltwater.

8. (1 pt) <u>They tell us that the earth can't be billions of years old</u>. If the earth was more than a few million years old, the oceans would be much saltier.

9. (2 pts – one for the fact that it does not and one for the why) <u>It does not form an iceberg. Icebergs are freshwater and come from glaciers</u>. Sea ice has salt mixed in with it.

10. (1 pt) <u>Precipitation</u> is responsible for glaciers. Remember, glaciers start because of snow, and snow is precipitation.

11. (1 pt) <u>Adiabatic cooling</u> causes the temperature change that allows for the condensation that makes most clouds.

12. (1 pt) <u>The temperature will increase</u>. Gases cool as they expand and heat up as they are compressed if they are not allowed to exchange energy with their surroundings.

13. (1 pt) <u>The depth of the water table will decrease.</u> If a lot more water enters the soil, more soil than usual will become saturated. This means you don't have to go down as far to find the saturated soil, so the water table is no longer as deep.

14. (1 pt) <u>The nature of groundwater flow makes it such that a lake can be polluted by groundwater that originally soaked into the soil hundreds of miles away.</u> When you find a lake polluted by groundwater pollution, how will you know where it came from?

Total possible points: 20

SOLUTIONS TO THE TEST FOR MODULE #6

1. (4 pts – one for each definition)

a. <u>Sedimentary rock</u> – Rock formed when chemical reactions cement sediments together, hardening them

b. <u>Plastic rock</u> - Rock that behaves like something between a liquid and a solid

c. <u>Fault</u> – The boundary between two sections of rock that can move relative to one another

d. <u>Epicenter</u> – The point on the surface of the earth directly above an earthquake's focus

2. (5 pts – one for each region)

    a. <u>crust</u>
    b. <u>asthenosphere</u>
    c. <u>mantle</u>
    d. <u>outer core</u>
    e. <u>inner core</u>

3. (1 pt) Scientists have observed <u>seismic waves</u>. The behavior of these waves tells us a lot about the makeup and properties of the mantle and core. The student could say "sound waves" or just "waves" instead of "seismic waves."

4. (1 pt – ½ for each region) The Moho separates the <u>crust</u> from the <u>mantle.</u>

5. (1 pt) <u>Electrical flow in the core</u> causes the earth's magnetic field.

6. (2 pts – ½ for each theory and one for which is the most valid) <u>The dynamo theory and the rapid-decay theory</u> both attempt to explain the earth's magnetic field. <u>The rapid-decay theory is more scientifically valid.</u>

7. (1 pt) <u>The earth's magnetic field blocks cosmic rays from the sun.</u>

8. (1 pt) <u>The trench is probably the site where one plate interacts with another.</u> The student could discuss the specific interaction (like subduction or separation), but that is not necessary.

9. (1 pt) <u>The region nearest the plate boundary should have more earthquakes.</u>

10. (1 pt) Each step on the Richter scale means a factor of 32 in energy. Since the quake and aftershock are off by 2 units, the quake was 32x32 = <u>1,024 times more energetic than the aftershock.</u>

11. (1 pt) <u>You expect to find both volcanic mountains and domed mountains.</u>

12. (1 pt) Scientists typically call this supercontinent <u>Pangaea.</u>

Total possible points: 20

SOLUTIONS TO THE TEST FOR MODULE #7

1. (5 pts – one for each definition)

a. Aphelion – The point at which the earth is farthest from the sun

b. Perihelion – The point at which the earth is closest to the sun

c. Coriolis effect – The way in which the rotation of the earth bends the path of winds, sea currents, and objects that fly through different latitudes

d. Air mass – A large body of air with relatively uniform pressure, temperature, and humidity

e. Weather front – A boundary between two air masses

2. (3 pts – one for each cloud type) a. stratus (stratonimbus or nimbostratus is okay also)
b. cirrus   c. cumulus

3. (1 pt) The three main factors are thermal energy, uneven distribution of thermal energy, and water vapor in the atmosphere. Winds are caused by imbalances of temperature, which is due to uneven distribution of thermal energy.

4. (1 pt) Dark clouds have a "nimbo" prefix or a "nimbus" suffix. The proper term is cumulonimbus, but nimbocumulus works as well.

5. (1 pt) Insolation stands for incoming solar radiation, which is sunlight. If a region gets a lot of sunlight, it is warm.

6. (2 pts – one for each answer) From the winter solstice to the summer solstice, day times in the Northern Hemisphere increase, because the Northern Hemisphere begins pointing toward from the sun. At the spring equinox, the day length is 12 hours. Since the day time is increasing after the winter solstice, the day times must still be less than 12 hours from the winter solstice to the spring equinox.

7. (1 pt) The Northern Hemisphere is pointed away from the sun at perihelion. This is a much greater effect than the distance from the sun.

8. (1 pt) You would fire it northwest to counteract the Coriolis effect. See Figure 7.8.

9. (2 pts – one for each answer) The change in temperature caused by air changing latitude, along with the Coriolis effect, cause the global wind patterns that we see on the earth.

10. (1 pt) Local winds interfere with the global wind patterns.

11. (1 pt) This is a continental tropical air mass.

12. (1 pt) This cloud progression is typical of a warm front, which causes long, lighter rain.

Total possible points: 20

SOLUTIONS TO THE TEST FOR MODULE #8

1. (2 pts – one for each definition)

a. <u>Updraft</u> – A current of rising air

b. <u>Insulator</u> – A substance that does not conduct electricity very well

2. (1 pt) <u>The Bergeron process</u> is at work here, because that's the process that says the precipitation starts out as snow and melts into rain as it reaches warm air.

3. (3 pts – one for the definition and one each for the two influences) <u>The dew point is the temperature at which water vapor condenses out of the air onto ground-level surfaces.</u> It is influenced by <u>humidity</u> and <u>pressure</u>.

4. (2 pts – one for the stage and one for the fact that no hail will exist) <u>The thunderstorm is in its dissipation stage. No hail will exist</u> because hail must have an updraft to form.

5. (1 pt) <u>These thunderstorms are made up of several cells,</u> each of which produces heavy rains for about 30 minutes or less.

6. (1 pt) <u>The charge imbalance that causes lightning starts in a cloud and cannot form unless the cloud is tall</u>. Remember, the charge imbalance comes from the millions of collisions that take place between the falling raindrop or ice crystal and other things in the cloud. Without a tall cloud, there will not be enough collisions to cause the charge imbalance.

7. (1 pt) <u>The stepped leader</u> forms first.

8. (1 pt) <u>Lightning causes thunder by heating up the air through which it passes.</u> That heat generates a wave that we detect as sound.

9. (1 pt) <u>Sheet lightning cannot strike a person</u> because it is cloud-to-cloud lightning.

10. (1 pt) <u>Yes, it has touched ground.</u> That marks the beginning of the organization stage.

11. (1 pt) <u>Wind speed</u> differentiates all of the classifications of potential hurricanes.

12. (1 pt) The <u>eye</u> is calm.

13. (1 pt) The pressure <u>is equivalent</u> because the two cities are on the same isobar.

14. (1 pt) The pressure in Chicago is <u>higher</u> because Chicago is more than 3 isobars from the "L," while New York is only between 2 and 3 isobars from the "L."

15. (1 pt) <u>New York</u> might be experiencing thunderstorms because of the cold front.

16. (1 pt) <u>Houston</u> should expect warmer weather, since it is in the path of a warm front.

Total possible points: 20

SOLUTIONS TO THE TEST FOR MODULE #9

1. (5 pts – one for each definition)

a. <u>Reference point</u> – A point against which position is measured

b. <u>Vector quantity</u> – A physical measurement that contains directional information

c. <u>Scalar quantity</u> – A physical measurement that does not contain directional information

d. <u>Acceleration</u> – The time rate of change of an object's velocity

e. <u>Free fall</u> – The motion of an object when it is falling solely under the influence of gravity

2. (1 pt) In order for motion to occur, an object's position must change. In order to determine position, there must be a reference point. <u>The reference point allows you to determine whether or not position changes.</u> Note: the statement "all motion is relative" deserves ½ point.

3. a. (1 pt) <u>Your grandmother is in motion relative to you.</u> Even though your grandmother is standing still, her position relative to you is changing. Thus, she is in motion relative to you.

b. (1 pt) <u>Your mother is motionless relative to you.</u> Her position relative to you does not change. She is therefore motionless with respect to you.

4. (2 pts – one for the conversion to hours and one for the speed) This problem gives us distance and time and asks for speed. We know it is asking for speed because a distance unit divided by a time unit is speed or velocity. There is no direction here, so we are talking about speed. Thus, we need to use Equation (9.1). The problem wants the answer in miles per hour, however. We are given the time in minutes. Thus, we must make a conversion first:

$$\frac{30 \, \cancel{\text{minutes}}}{1} \times \frac{1 \, \text{hour}}{60 \, \cancel{\text{minutes}}} = 0.5 \, \text{hours}$$

Now we can use our speed equation:

$$\text{speed} = \frac{40 \, \text{miles}}{0.5 \, \text{hours}} = 80 \, \underline{\frac{\text{miles}}{\text{hour}}}$$

5. (2 pts – ½ for the conversion to seconds, ½ for the conversion to meters, and one for the speed) The problem wants velocity, which is speed and direction. To get speed, we will use Equation (9.1). Unfortunately, the problem tells us to give the answer in meters per second, but the distance is in kilometers and the time is in minutes. Thus, we need to do two conversions:

$$\frac{4.1 \, \cancel{\text{minutes}}}{1} \times \frac{60 \, \text{seconds}}{1 \, \cancel{\text{minute}}} = 246 \, \text{seconds}$$

$$\frac{1 \text{ km}}{1} \times \frac{1,000 \text{ m}}{1 \text{ km}} = 1,000 \text{ m}$$

Now we can use our speed equation:

$$\text{speed} = \frac{1,000 \text{ m}}{246 \text{ seconds}} = 4.07 \frac{\text{meters}}{\text{second}}$$

That's not quite the answer. The problem wants velocity, which includes direction. Thus, the answer is 4.07 meters/second west. Note that the student can have more decimal places in his answer.

6. a. (1 pt) This measurement has a distance unit divided by a time unit. That's speed or velocity. Since no direction is given, this is speed.

b. (1 pt) The unit of feet by itself measures distance.

c. (1 pt) This measurement has a distance unit divided by a time unit. That's speed or velocity. Since a direction is given, this is velocity.

d. (1 pt) This measurement has a distance unit divided by a time unit squared. That's acceleration. The direction is necessary because acceleration is a vector quantity.

7. (2 pts – one for the speed and one for the direction) As the picture shows, the eagle is behind the rabbit, but they are both traveling in the same direction. Thus, we get their relative velocity by subtracting their individual velocities:

$$\text{relative velocity} = 5.4 \text{ meters/second} - 4.4 \text{ meters/second} = 1.0 \text{ meter/second}$$

Since the rabbit is traveling faster than the eagle, the rabbit is pulling away. Thus, the relative velocity is 1.0 meter per second away from each other.

8. (1 pt) Since the velocity is not changing, the acceleration is zero.

9. (2 pts – one for the value and one for the direction) The initial velocity is 12 meters per second north, and the final velocity is 0. The time is 3 seconds. This is a straightforward application of Equation (9.2).

$$\text{acceleration} = \frac{\text{final velocity} - \text{initial velocity}}{\text{time}}$$

$$\text{acceleration} = \frac{0 \frac{\text{meters}}{\text{second}} - 12 \frac{\text{meters}}{\text{second}}}{3 \text{ seconds}} = \frac{-12 \frac{\text{meters}}{\text{second}}}{3 \text{ seconds}} = -4 \frac{\text{meters}}{\text{second}^2}$$

The negative just tells us that the acceleration is in the opposite direction of velocity. Thus, the acceleration is 4 meters/second^2 south.

10. (3 pts – one for the conversion, one for the value, and one for the direction) This is another application of Equation (9.2), because we are given time (5 seconds), initial velocity (0) and final velocity (60 miles per hour north). We can't use the equation yet, however, because our time units do not agree. We'll fix that first:

$$\frac{5 \text{ \cancel{seconds}}}{1} \times \frac{1 \text{ hour}}{3,600 \text{ \cancel{seconds}}} = 0.00139 \text{ hours}$$

Now that we have all time units in agreement, we can use the acceleration equation:

$$\text{acceleration} = \frac{\text{final velocity} - \text{initial velocity}}{\text{time}}$$

$$\text{acceleration} = \frac{60 \frac{\text{miles}}{\text{hour}} - 0 \frac{\text{miles}}{\text{hour}}}{0.00139 \text{ hours}} = \frac{60 \frac{\text{miles}}{\text{hour}}}{0.00139 \text{ hours}} = 43,165 \frac{\text{miles}}{\text{hour}^2}$$

The car speeds up, so acceleration is in the same direction as velocity. The answer, then, is 43,165 miles/hour² north. Note that the student's answer can have a different number of decimal places and can even be a slightly different value, depending on where and when the student rounded the numbers.

11. (1 pt) Neither has greater acceleration. Both objects are falling near the surface of the earth; thus, they are each in free fall. That means they both have equal acceleration. The ball was given more *initial velocity*, so it will travel faster. The acceleration on both is the same, however.

12. (1 pt) The feather is more affected by air resistance than the rock. This is the same situation as Experiment 9.2.

13. (2 pts – one for setting up the equation properly and one for getting the right answer) The rock is in free fall, so we can use Equation (9.3). Since the problem wants the answer in feet, we need to use 32 feet per second² as the acceleration.

$$\text{distance} = \frac{1}{2} \times (\text{acceleration}) \times (\text{time})^2$$

$$\text{distance} = \frac{1}{2} \times (32 \frac{\text{feet}}{\text{second}^2}) \times (1.3 \text{ seconds})^2 = \frac{1}{2} \times (32 \frac{\text{feet}}{\text{second}^2}) \times (1.3 \text{ seconds}) \times (1.3 \text{ seconds})$$

$$\text{distance} = \frac{1}{2} \times (32 \frac{\text{feet}}{\cancel{\text{second}^2}}) \times (1.69 \text{ \cancel{seconds}}^2) = \underline{27.04 \text{ feet}}$$

Note that the student can have a different number of decimal places in his answer.

Total possible points: 28

SOLUTIONS TO THE TEST FOR MODULE #10

1. (4 pts – one for each definition)

a. <u>Inertia</u> – The tendency of an object to resist changes in its velocity

b. <u>Friction</u> – A force that opposes motion, resulting from the contact of two surfaces

c. <u>Kinetic friction</u> – Friction that opposes motion once the motion has already started

d. <u>Static friction</u> – Friction that opposes the initiation of motion

2. (3 pts – one for each law)

<u>Newton's First Law</u> – An object in motion (or at rest) will tend to stay in motion (or at rest) until it is acted upon by an outside force.

<u>Newton's Second Law</u> – When an object is acted on by one or more outside forces, the total force is equal to the mass of the object times the resulting acceleration

<u>Newton's Third Law</u> – For every action, there is an equal and opposite reaction.

3. (1 pt) <u>The pilot must drop the bombs before the plane reaches the airfield.</u> The bombs will have a velocity equal to that of the plane when they are dropped. Thus, they will continue to fly north as they fall. In order to hit the airfield, then, they must be dropped south of it.

4. (1 pt) <u>The mouse will start moving straight, skidding off of the disk.</u> This is like Experiment 10.2. When the disk stops, the mouse has a velocity pointed in a certain direction. Without sufficient time, the frictional force will not be able to keep the mouse on the disk. This will cause the mouse to start traveling in a straight line, in the direction it was moving right before the disk stopped.

5. (1 pt) Since we are ignoring friction, this is an easy problem:

$$\text{total force} = (\text{mass}) \cdot (\text{acceleration})$$

$$\text{total force} = (1.0 \, \text{kg}) \cdot \left(4.0 \, \frac{\text{m}}{\text{sec}^2} \right) = 4.0 \, \text{Newtons}$$

The cube is being pulled with a force of <u>4.0 Newtons down the tray</u>.

6. (1 pt) The baseball player is slowing, because his velocity is north but the acceleration is directed south. Friction slows things down. This is the only force in the problem, since nothing else is pulling or pushing on the player. Thus, the force that results from the acceleration will be the frictional force.

$$\text{total force} = (\text{mass}) \cdot (\text{acceleration})$$

$$\text{total force} = (75 \text{ kg}) \cdot \left(5.0 \frac{\text{m}}{\text{sec}^2} \right) = 375 \text{ Newtons}$$

The frictional force is 375 Newtons to the south.

7. (2 pts – one for the force and one for the fact that it is kinetic) If the cart is moving with a constant velocity, that means the acceleration (and thus the total force) equals zero. Thus, the man must be pushing with just enough force to counteract friction. Thus, the frictional force is 200 Newtons against the motion of the cart. This is kinetic friction, because the cart is moving.

8. (1 pt) Since static friction is generally greater than kinetic friction, 11 Newtons refers to static friction, and 8 Newtons refers to kinetic friction.

9. (3 pts – one for the force to get the box moving, one for the total force, and one for the answer) If the static frictional force is 20 Newtons, the woman must overcome it to get the box moving. To get the box moving, then, a force greater than 20 Newtons must be used. Once the box is moving, the acceleration is 1.0 meters per second^2, so the total force is:

$$\text{total force} = (\text{mass}) \cdot (\text{acceleration})$$

$$\text{total force} = (30 \text{ kg}) \cdot \left(1.0 \frac{\text{m}}{\text{sec}^2} \right) = 30 \text{ Newtons}$$

This force is the combination of the woman's force and the kinetic frictional force (7 Newtons). Since the kinetic frictional force opposes motion, it is opposite of the woman's force. Thus, the total force is the woman's force minus the frictional force.

$$\text{woman's force} - 7 \text{ Newtons} = 30 \text{ Newtons}$$

Thus, in order for the total force to be 30 Newtons, the woman's force must be 37 Newtons west.

10. (1 pt) The ball exerts a force on the bat. This is the equal and opposite force demanded by Newton's Third Law.

Total possible points: 18

SOLUTIONS TO THE TEST FOR MODULE #11

1. (2 pts – one for each arrow) In circular motion, the velocity is always straight in the direction of travel, and the force is a centripetal force that points toward the center of the circle.

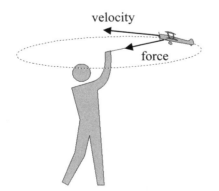

2. (1 pt) The weakest of the fundamental forces is the gravitational force.

3. (1 pt) The ball exerts an equal and opposite force on the earth.

4. (2 pts – ½ for the effect of each mass change and one for the answer) When the 5-kg mass is replaced by a 20-kg mass, the mass is multiplied by 4. This multiplies the gravitational force by 4. The 2-kg mass is multiplied by 6 when it is replaced by the 12-kg mass, so that further multiplies the force by 6. Thus, the total change is 4x6 = 24. The new gravitational force, then, is 24 times larger than the old one.

5. (2 pts – one for seeing the factor by which distance was changed and one for the answer) The only difference is that the distance between the objects was multiplied by 3. The gravitational force decreases when the distance between the objects increases. It is decreased according to the square of that increase. Thus, the force is divided by 3^2, which is 9. The new gravitational force, then, is 9 times smaller than the old one.

6. (1 pt) The first one moves faster, because it takes less time to make the orbit. The principles of circular motion state that the faster the speed, the greater the centripetal force necessary. Gravity supplies the centripetal force in an orbit. Thus, the first moon experiences the greatest gravitational force.

7. (2 pts – ½ for each) Jupiter, Saturn, Uranus, and Neptune

8. (1 pt) The planet farthest away from the sun receives the least insolation. Thus, the answer is Neptune.

9. (1 pt) Comet orbits are usually highly elliptical, and the closer they get to the sun, the easier they are to see. Thus, the answer is c.

10. (1 pt – ½ for each) The coma and tail come and go, depending on the proximity to the sun.

11. (1 pt) <u>The Kuiper belt is thought to be a source of short-period comets</u>. An answer of just "comets" receives only half credit.

12. (1 pt) <u>The graviton theory</u> states that gravity is caused by the exchange of particles.

13. (1 pt) <u>The General Theory of Relativity</u> would be true, because this theory states that gravity is not actually a force. Instead, it is simply a consequence of how mass bends space and time.

Total possible points: 17

SOLUTIONS TO THE TEST FOR MODULE #12

1 (7 pts – one for each definition)

a. Photon – A small "package" of light that acts like a particle

b. Charging by conduction – Charging an object by allowing it to come into contact with an object that already has an electrical charge

c. Charging by induction – Charging an object without direct contact between the object and a charge

d. Electrical current – The amount of charge that travels past a fixed point in an electric circuit each second

e. Conventional current – Current that flows from the positive side of the battery to the negative side. This is the way current is drawn in circuit diagrams, even though it is wrong.

f. Resistance – The ability of a material to impede the flow of charge

g. Open circuit – A circuit that does not have a complete connection between the two sides of the power source. As a result, current does not flow.

2. (1 pt) Opposite charges attract one another. Thus, each will pull the other to itself.

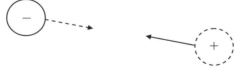

3. (2 pts – one for the factor and the other for the fact that it is repulsive) The electromagnetic force varies inversely with the square of the distance between magnets. Thus, if the distance is doubled, the force decreases by a factor of 4. Since the poles are the same, this is a repulsive force.

4. (1 pt) The object developed a charge opposite of the rod. Thus, the object was charged by induction.

5. (1 pt – ½ for the fact that you can and ½ for how) A low voltage means that each electron will have only a little bit of energy. You can still get a lot of energy out of the circuit, however, as long as you put a lot of current through it.

6. (1 pt) The cord in your right hand is thicker. The thicker the cord, the lower the resistance. The lower the resistance, the less heat will be made.

7. (1 pt) Conventional current runs from the positive end of the battery (the large line) to the negative end (the small line). The electron flow is opposite of that.

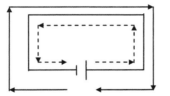

8. (1 pt) <u>The light bulbs are wired in parallel</u>.

9. (1 pt) <u>The object is not a magnet.</u>

10. (1 pt) <u>You will have two north poles and two south poles</u>. As soon as the magnet is cut, each pole will change. Since it is impossible to have just one pole of a magnet, the north pole will change into a magnet with a north and south pole. The south pole will also change into a magnet with a north and south pole. Thus, you will have 2 magnets, each with a north and south pole.

Total possible points: 17

SOLUTIONS TO THE TEST FOR MODULE #13

1. (6 pts – one for each definition)

a. <u>Nucleus</u> – The center of an atom, containing the protons and neutrons

b. <u>Atomic number</u> – The number of protons in an atom

c. <u>Mass number</u> – The sum of the numbers of neutrons and protons in the nucleus of an atom

d. <u>Isotopes</u> – Atoms with the same number of protons but different numbers of neutrons

e. <u>Element</u> – A collection of atoms that all have the same number of protons

f. <u>Radioactive isotope</u> – An atom with a nucleus that is not stable

2. (2 pts – one for each answer) A <u>proton</u> is larger than an electron. An <u>electron</u> is negatively charged.

3. (1 pt – ½ for each answer) The nuclear force is caused by <u>the exchange of pions</u>. It is a short-range force because <u>pions can only exist for a short time</u>.

4. a. (1 pt – one-third for each answer) Since the chemical symbol is Ca, we can use the chart to learn that the atom has <u>20 protons</u>. This tells us there are also <u>20 electrons</u>. The mass number is the sum of protons and neutrons in the nucleus. Thus, there are <u>28 neutrons</u>.

b. (1 pt – one-third for each answer) Since the chemical symbol is Sn, we can use the chart to learn that the atom has <u>50 protons</u>. This tells us there are also <u>50 electrons</u>. The mass number is the sum or protons and neutrons in the nucleus. Thus, there are <u>74 neutrons</u>.

c. (1 pt – one-third for each answer) Since the chemical symbol is Ag, we can use the chart to learn that the atom has <u>47 protons</u>. This tells us there are also <u>47 electrons</u>. The mass number is the sum of protons and neutrons in the nucleus. Thus, there are <u>62 neutrons</u>.

5. (2 pts – ½ for the correct number of protons, ½ for the correct number of neutrons, and one for the correct arrangement of the electrons) All atoms symbolized with "Na" have 11 protons according to the chart. This also means there are 11 electrons. Two of them can go into the first Bohr orbit, and 8 can go in the second Bohr orbit. We will have to put the remaining one in the third Bohr orbit. The mass number indicates that there are 12 neutrons:

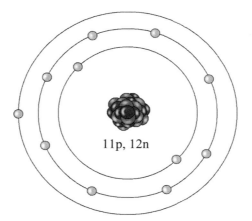
11p, 12n

6. (1 pt – ½ for each. For every wrong answer, take ½ off, but don't allow the value to be negative.) Isotopes have the same number of protons but different numbers of neutrons. Since chemical symbols tell us the number of protons, we are looking for atoms with the same chemical symbol but different mass numbers. Thus, ^{144}Nd and ^{145}Nd are isotopes.

7. (1 pt) In the first three hours, the 30 gram sample will be reduced to 15 grams. In the next three hours, it will be reduced to 7.5 grams. In the next 3 hours, it will be reduced to 3.75 grams. In the next three hours, it will be reduced to 1.875 grams. In the next 3 hours, it will be reduced to 0.9375 grams.

8. (1 pt) ^{144}Ce has 58 protons according to the chart. This means there must be 86 neutrons. In beta decay, a neutron turns into a proton. This will result in an atom with 59 protons and 85 neutrons, or ^{144}Pr.

9. (1 pt) ^{220}Rn has 86 protons according to the chart. This means there must 134 neutrons. In alpha decay, the nucleus loses 2 protons and 2 neutrons. This will result in an atom with 84 protons and 132 neutrons, or ^{216}Po.

10. (1 pt) In gamma decay, the number of protons and neutrons is unaffected by the decay. Thus, the daughter product is the same as the original ^{239}U.

11. (1 pt) The person would only be protected from alpha particles.

Total possible points: 20

SOLUTIONS TO THE TEST FOR MODULE #14

1. (5 pts – one for each definition)

a. Transverse wave – A wave with a direction of propagation that is perpendicular to its direction of oscillation

b. Longitudinal wave – A wave with a direction of propagation that is parallel to its direction of oscillation

c. Supersonic speed – Any speed that is faster than the speed of sound in the substance of interest

d. Sonic boom – The sound produced as a result of an object traveling at or above Mach 1

e. Pitch – An indication of how high or low a sound is, which is primarily determined by the frequency of the sound wave

2. (1 pt) Low pitch means low frequency. Since wavelength and frequency are inversely proportional, low frequency means long wavelength. Thus, the second recorder (the longer one) produces the lower notes.

3. (1 pt) No one can hear you scream in space because there is little to no air in space. As a result, sound waves cannot travel because there is no medium for them to oscillate.

4. (1 pt) Sound waves are longitudinal waves. This means they oscillate parallel to the direction in which they travel.

5. (1 pt) The speed of sound is given by Equation (14.2):

$$v = (331.5 + 0.6 \cdot 25)\,\frac{m}{sec}$$

$$v = (331.5 + 15)\,\frac{m}{sec}$$

$$v = 346.5\,\frac{m}{sec}$$

6. (2 pts – one for the speed and one for the answer) To determine frequency, we need to use Equation (14.1). To use that equation, however, we need to know the speed and wavelength. We are given the wavelength (2 m), and we can get the speed from the temperature and Equation (14.2):

$$v = (331.5 + 0.6 \cdot 17)\,\frac{m}{sec}$$

$$v = (331.5 + 10.2)\ \frac{m}{sec}$$

$$v = 341.7\ \frac{m}{sec}$$

Now that we have the speed, we can finally use Equation (14.1):

$$f = \frac{v}{\lambda}$$

$$f = \frac{341.7\ \frac{m}{sec}}{2\ m} = 170.85\ \frac{1}{sec}$$

The frequency is <u>170.85 Hz</u>.

7. (1 pt) Infrasonic waves have the lowest frequencies; sonic waves have higher frequencies; and ultrasonic waves have the highest frequencies. Since frequency and wavelength are inversely proportional, <u>infrasonic waves</u> have the longest wavelengths.

8. (2 pts – one for the speed and one for the answer) To determine the distance, we will use the time difference between the lightning flash and the sound. We will assume that the light from the lightning reaches your eyes essentially at the same time as the lightning was formed. Thus, the time it takes for the sound to travel to you will determine the distance. First, then, we need to know the speed of sound:

$$v = (331.5 + 0.6 \cdot 10)\ \frac{m}{sec}$$

$$v = (331.5 + 6)\ \frac{m}{sec}$$

$$v = 337.5\ \frac{m}{sec}$$

Now we can use Equation (14.3):

$$\text{distance traveled} = (\text{speed}) \times (\text{time traveled})$$

$$\text{distance traveled} = (337.5\ \frac{m}{sec}) \times (2\ sec) = \underline{675\ m}$$

9. (1 pt) Sound waves travel faster in liquids than they do in gases. Thus, the sound will travel faster in <u>water</u>.

10. (2 pts – ½ for each answer) Since the man is singing the low notes, the man's sound waves have the <u>lower frequencies</u> and the <u>longer wavelengths</u>. The speed of sound depends only on the medium and temperature. Thus, <u>the speed of the man's sound waves is the same as that of the woman's sound waves</u>. Finally, amplitude determines loudness. Therefore, <u>the amplitude of the man's sound waves is smaller</u>.

11. (2 pts – one for the speed of sound and one for the answer) To determine the speed of the jet, we first have to determine the speed of sound. After all, Mach 3 means 3 times the speed of sound. Thus, we need to know the speed of sound in order to determine the speed of the jet.

$$v = (331.5 + 0.6 \cdot 10) \frac{m}{sec}$$

$$v = (331.5 + 6) \frac{m}{sec}$$

$$v = 337.5 \frac{m}{sec}$$

Since sound travels at 337.5 m/sec, Mach 3 is 3 x (337.5 m/sec) = <u>1,012.5 m/sec</u>.

12. (1 pt) Since the pitch of the siren gets higher, you are moving so that the wavelength of the sound waves seems smaller. This means you are moving <u>toward</u> the siren, because that will cause you to encounter the crests of the sound waves faster than you would if you were still.

13. (2 pts – ½ for each conversion and one for the answer) The bel scale states that every bel unit corresponds to a factor of ten in the intensity of the sound waves. Thus, we need to determine how many bel units the sound was before and after the amplifier:

$$\frac{30 \cancel{decibels}}{1} \times \frac{0.1 \, bel}{1 \, \cancel{decibel}} = 3 \, bels$$

$$\frac{80 \cancel{decibels}}{1} \times \frac{0.1 \, bel}{1 \, \cancel{decibel}} = 8 \, bels$$

Since the sound is 5 bels louder after the amplifier, the increase in sound wave intensity is 5 factors of ten higher. Thus, the sound wave intensity is 10 x 10 x 10 x 10 x 10 = <u>100,000 times larger after the amplifier as compared to before</u>.

Total possible points: 22

SOLUTIONS TO THE TEST FOR MODULE #15

1. (2 pts – one for each definition)

a. <u>Electromagnetic wave</u> – A transverse wave composed of an oscillating electric field and a magnetic field that oscillates perpendicular to the electric field

b. <u>The Law of Reflection</u> – The angle of reflection equals the angle of incidence.

2. (3 pts – one for the idea of "packets" or "particles," one for the idea of a transverse wave, and one for the fact that there are both electric and magnetic oscillations) <u>Light is made up of individual packets. Each packet is made up of two perpendicular transverse waves: one that consists of an oscillating electrical field, and one that is made up of an oscillating magnetic field.</u>

3. (1 pt) Light travels more slowly in solids than gases. Thus, the light will <u>slow down</u> when it enters the glass.

4. (1 pt) Yellow light has a longer wavelength than green light. Thus, <u>green light</u> has a higher frequency.

5. (1 pt) <u>Radio waves</u> have a longer wavelengths.

6. (3 pts – one for the first refraction into the aquarium, one for the reflection on the right side, and one for the refraction on the right side) The light ray will bend towards the perpendicular when it refracts into the aquarium. There will also be a reflected ray there, but that does not affect the answer to this problem. When it reaches the other side, part of it will refract out and bend away from the perpendicular. The rest will reflect away. To get full credit, the student must show the ray leaving the aquarium and bending away from the perpendicular, and the student must also show the reflected ray with an angle equal to that of the incident ray.

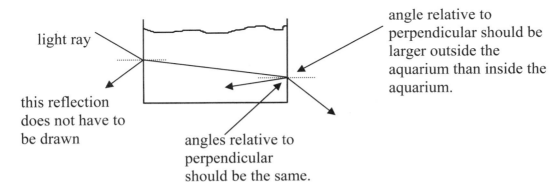

light ray

this reflection does not have to be drawn

angles relative to perpendicular should be the same.

angle relative to perpendicular should be larger outside the aquarium than inside the aquarium.

7. (1 pt) The quarter is not really 3 feet in front of you because <u>as the light from the quarter leaves the pool water, it bends. This causes a false image.</u> The quarter is really a bit closer to you.

8. (1 pt) You would use a <u>converging lens</u>. A diverging lens would cause the light rays to move away from each other, not converge to a single point.

9. (1 pt) To change its focus, the eye <u>changes the shape of the lens</u>.

For problems 10-12, remember that blue + green = cyan, red + blue = magenta, and red + green = yellow.

10. (1 pt) Since a computer screen shines light at your eyes, the colors add. Thus, you get a <u>cyan</u> dot.

11. (1 pt) Cyan ink absorbs all colors except blue and green. Thus, when white light shines on a cyan spot, only blue and green light reflect off it. That's what gives it the cyan color. In the same way, magenta ink absorbs all colors except blue and red. Therefore, a mixture of the two will absorb all colors but blue. As a result, the ink will be <u>blue</u>.

12. (1 pt) Yellow absorbs all colors except red and green. Thus, the red light will be reflected back. That will be the *only* color reflected back, so it will appear <u>red</u>.

Total possible points: 17

SOLUTIONS TO THE TEST FOR MODULE #16

1. (7 pts – one for each definition)

a. <u>Nuclear fusion</u> – The process by which two or more small nuclei fuse to make a bigger nucleus

b. <u>Nuclear fission</u> – The process by which a large nucleus is split into smaller nuclei

c. <u>Critical mass</u> – The amount of isotope necessary to sustain a chain reaction

d. <u>Absolute magnitude</u> – The brightness of a star, corrected for distance, on a scale of -8 to +19. The *smaller* the number, the *brighter* the star.

e. <u>Apparent magnitude</u> – The brightness of a star as seen in the night sky. The *smaller* the number, the *brighter* the star.

f. <u>Light year</u> – The distance light could travel along a straight line in one year

g. <u>Galaxy</u> – A large ensemble of stars, all interacting through the gravitational force and orbiting around a common center

2. (1 pt) <u>Nuclear fusion</u> occurs in the sun's core.

3. (1 pt) <u>This is nuclear fusion</u>, since two small nuclei fused to make a bigger one.

4. (1 pt) He would get a <u>positive number</u>. In nuclear fission and fusion, mass is converted into energy. Thus, the mass of the materials produced will be less than the mass of the starting materials.

5. (2 pts – one for the fact that it is not possible and one for the why) <u>It is not possible, because a power plant does not have significantly more than the critical mass of the large nucleus.</u>

6. (4 pts – one for each letter) Using the magnitudes and spectral letters given, we have:

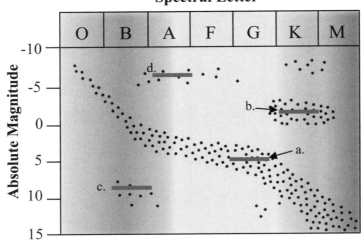

This tells us the classifications are:

a. <u>Main sequence</u>   b. <u>Red giant</u>   c. <u>White dwarf</u>   d. <u>Supergiant</u>

7. (1 pt) The brightest star is the one with the lowest magnitude. Thus, <u>(d)</u> is the brightest star.

8. (1 pt) The hottest star is the one that is the farthest to the left on the H-R diagram, because temperature increases the farther left you travel on the H-R diagram. Thus, <u>(c)</u> is the hottest star.

9. (1 pt – ½ for each) The two main types of variable stars are <u>novas</u> and <u>pulsating variables</u>.

10. (2 pts – one for what and one for why) <u>Cepheid variables are variable stars whose magnitude and period have a direct relationship. They are important in astronomy because that relationship can be used to measure long distances in the universe.</u>

11. (2 pts – one for each) Earth's solar system belongs to the <u>Milky Way</u> galaxy, which is a <u>spiral galaxy</u>.

12. (1 pt) <u>The red shift is the phenomenon in which light that comes to the earth from other galaxies ends up having longer wavelengths than it should.</u> Most astronomers interpret that as a Doppler shift resulting from the expansion of the universe.

Total possible points: 24

QUARTERLY TEST #1

1. Define the following terms:

a. Molecule
b. Relative humidity
c. Barometer
d. Jet streams
e. Solvent
f. Cohesion

2. If you wanted to measure an object's volume, what metric unit would you use? What English unit would you use?

3. How many centimeters are in 12 meters?

4. If an object has a mass of 75 g, what is its mass in kg?

5. How many gallons of milk are in 0.5 liters of milk? (1 gal = 3.78 L)

6. Two different cleaning products have exactly the same active ingredient. One is "industrial strength" and cleans much more effectively than the other, which is "regular strength." If they each have exactly the same active ingredient, what is the difference between the two cleaners?

7. What is the greenhouse effect and why is it important for life on earth?

8. If you put a glass of water outside when the relative humidity is 100%, how quickly will the water evaporate?

9. Why does sweating cool people down?

10. What makes up the majority of the air that we inhale?

11. Do we exhale more carbon dioxide or more oxygen?

12. Is the air cleaner today or was it cleaner 30 years ago?

13. Why do we have laws limiting the amount of ozone that can be produced by industrial plants and automobiles? Since ozone protects us from the ultraviolet rays of the sun, why don't we want as much ozone produced as possible?

14. The average, sea-level value for atmospheric pressure is 14.7 pounds per square inch, which is the same as 29.9 inches of mercury. If the atmospheric pressure is 1.04 atms, which of the following values would correspond to atmospheric pressure as reported in a weather report?

31.1 inches of mercury, 29.9 inches of mercury, 25.4 inches of mercury

15. Two vials contain air samples taken at different altitudes. The first is composed of 21% oxygen, 78% nitrogen, and 1% other. The second is 95% helium, 4% hydrogen, and 1% other. Which came from the homosphere?

16. Name the three regions of the homosphere, from lowest to highest.

17. In which regions of the homosphere does temperature decrease with increasing altitude? In what region does temperature increase with increasing altitude?

18. In what region of the homosphere does most of the earth's weather occur?

19. You have two samples of liquid. The first has a significantly higher temperature than the second. In which sample do the molecules move more slowly?

20. Why is the "ozone hole" a seasonal event that is localized mostly over Antarctica?

21. The natural gas that you burn in a stove or furnace is composed primarily of methane, a molecule with one carbon atom (C), and four hydrogen atoms (H). What is the chemical formula of methane?

22. Plants manufacture glucose for food. If the chemical formula for glucose is $C_6H_{12}O_6$, how many atoms are in a molecule of glucose?

23. A molecule is composed of atoms that all pull on electrons with the same strength. Will this molecule be nonpolar?

24. Carbon disulfide will not dissolve in water. Is it made of ionic molecules, polar molecules, or nonpolar molecules?

25. What is responsible for the fact that, unlike virtually any other substance in the world, solid water (ice) is lighter than an equal volume of liquid water?

QUARTERLY TEST #2

1. Define the following terms:

a. Transpiration
b. Adiabatic cooling
c. Sedimentary rock
d. Aphelion
e. Perihelion
f. Coriolis effect
g. Insulator

2. A sample of water is taken randomly from earth's hydrosphere. What kind of water (freshwater or saltwater) is it most likely to be?

3. A lake has no outlets to dump into rivers or streams. The only way water can leave the lake is by evaporation. Does the lake contain freshwater or saltwater?

4. Why is the salinity of the ocean evidence that the earth is not billions of years old?

5. Where do glaciers come from?

6. If no energy is added to air, what happens to the temperature when the air expands?

7. Scientists often separate the earth into five distinct sections. Name those sections.

8. What is the main thing scientists observe in order to learn about the makeup of the earth's interior?

9. What causes the magnetic field of the earth?

10. Give a brief description of the two main theories that attempt to explain the earth's magnetic field. Which is more scientifically valid?

11. Why would life cease to exist without the earth's magnetic field?

12. What is Pangaea?

13. Name the four kinds of mountains.

14. What are the three main factors that affect earth's weather?

15. What does "insolation" stand for?

16. Is the Northern Hemisphere's summer during aphelion or perihelion?

17. What causes wind?

18. Over a period of a couple of days, the clouds slowly build and then a gentle, long rain ensues that lasts less than a full day. What kind of front causes this weather?

19. In less than a day, dark cumulonimbus clouds form and unleash a thunderstorm that lasts only a few hours. After the thunderstorm is over and the sky clears, do you expect cooler or warmer temperatures as compared to the temperature before the clouds began forming?

20. If the heavy rain of a thunderstorm lasts for more than 30 minutes, what can you conclude about its makeup?

21. Which is responsible for most of the light and sound in a lightning bolt: the stepped leader or the return stroke?

22. Where does the thunder in a thunderstorm come from?

23. What is the difference between sheet lightning and a lightning bolt?

24. What kind of cloud is necessary for tornado formation?

25. What causes a hurricane in the Southern Hemisphere to rotate in a different direction from a hurricane in the Northern Hemisphere?

QUARTERLY TEST #3

1. Define the following terms:

   a. Vector quantity
   b. Scalar quantity
   c. Acceleration

   d. Kinetic friction
   e. Static friction
   f. Photon

   g. Electrical current
   h. Resistance

2. What is the speed of a boat that travels 20 miles in 45 minutes? Please answer in miles per hour.

3. Label each quantity as a vector or scalar quantity. Also, identify it as speed, distance, velocity, acceleration, or none of these.

   a. 10 meters/second^2 north
   b. 1.2 meters/second
   c. 3.4 feet/hour and slowing
   d. 2.3 miles/minute west

4. A car goes from a velocity zero to a velocity of 15 meters per second east in 2.1 seconds. What is the car's acceleration?

5. What is the height of a building (in meters) if it takes a rock 3.8 seconds to drop from its roof?

6. A car and a truck are traveling north on a highway. The truck has a speed of 42 miles per hour and the car has a speed of 37 miles per hour. If the truck is ahead of the car, what is the relative velocity?

37 miles per hour north   42 miles per hour north

Illustrations from the MasterClips collection

7. If an object travels for 10 minutes with a constant velocity of 11 miles per hour north, what is the acceleration?

8. A boy is running north with a beanbag in his hands. He passes a tree and, at the moment he is beside the tree, he drops the beanbag. Will the beanbag land next to the tree? If not, will it be north or south of the tree?

9. A child is pushing her toy across the room with a constant velocity to the east. If the static friction between this toy and the floor is 17 Newtons, while the kinetic friction is 12 Newtons, what force is the child exerting?

10. A father is trying to teach his child to ice skate. As the child stands still, the father pushes him forward with an acceleration of 2.5 meters per second^2 north. If the child's mass is 22 kilograms, what is the force with which the father is pushing. (Since they are on ice, you can ignore friction.)

11. The static frictional force between a 500 kilogram box of bricks and the floor is 500 Newtons. The kinetic frictional force is only 220 Newtons. How many Newtons of force must a worker exert to get the box moving?

12. A man leans up against a wall with a force of 20 Newtons to the east. What is the force exerted by the wall on the man?

13. A physicist is measuring the frictional force between a box and the floor. The physicist writes down two numbers: 100 Newtons and 225 Newtons. Which is the static frictional force and which is the kinetic frictional force?

14. Name the four fundamental forces in creation. Which two forces are really different aspects of the same force?

15. Which is the weakest of the fundamental forces? Which is the strongest?

16. The gravitational force between two objects (mass$_1$ = 5 kg, mass$_2$ = 2 kg) is measured when the objects are 5 centimeters apart. If the distance between them is increased to 10 centimeters, how does the new gravitational attraction compare to the first one that was measured?

17. In the following diagram, the ball is traveling from point "A" to point "B." Draw the velocity of the ball and force it experiences if it is traveling at constant speed.

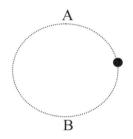

18. List the inner planets and the outer planets.

19. What are the three parts of a comet? Which of those parts is always present in a comet?

20. An electrical circuit uses a low voltage and a large current. Is the energy of each electron high or low? Are there many electrons flowing through the circuit or are there few?

21. Draw the conventional current flow in the following circuit.

22. You have two wires of the same length. One is thin, and the other is very thick. When the same current is run through each wire, which will heat up the fastest?

23. Three lights are in a room. When one burns out, the others remain lit, despite the fact that all light bulbs run on the same electrical circuit. Are the lights wired in a parallel circuit or a series circuit?

24. What causes a magnetic field?

25. Is it possible to have a permanent magnet that has only a south pole?

QUARTERLY TEST #4

1. Define the following terms:

a. Mass number	e. Longitudinal wave	i. Nuclear fusion
b. Isotopes	f. Supersonic speed	j. Nuclear fission
c. Atomic number	g. Electromagnetic wave	k. Light year
d. Transverse wave	h. The Law of Reflection	

2. Order the three constituent parts of the atom in terms of their size, from smallest to largest.

3. What force keeps the protons and neutrons in the nucleus? What causes this force?

4. List the number of protons, electrons and neutrons in a ^{14}C atom.

5. Determine the daughter product produced in the beta decay of ^{131}I.

6. The half-life of the radioactive decay of ^{226}Ra is 1,600 years. If a sample of ^{226}Ra originally had a mass of 10 grams, how many grams of ^{226}Ra would be left after 4,800 years?

7. List the three types of radioactive particles in the order of their ability to travel through matter. Start with the particle that cannot pass through much matter before stopping, and end with the one that can pass through the most matter before stopping.

8. The three classes of sound waves are sonic, infrasonic, and ultrasonic. Which class of sound waves has the shortest wavelengths?

9. An astronaut holds an alarm clock. When she puts on her space suit, she is able to hear the alarm when it rings. If the astronaut then walks out of her spaceship onto the surface of the moon, will the astronaut still be able to hear the alarm? Why or why not?

10. Are sound waves transverse waves or longitudinal waves?

11. You are watching the lightning from a thunderstorm. You suddenly see a flash of lightning and, 1.5 seconds later, you hear the thunder. How far away from you did the lightning strike? (The temperature at the time is 13 $^{\circ}$C).

12. Why do jets travel at speeds of Mach 1 or higher only in sparsely populated regions?

13. You hear two musical notes. They both have the same pitch, but the first is louder than the second. If you compared the sound waves of each sound, what aspect(s) of the waves (wavelength, frequency, speed, and amplitude) would be the same? What aspect(s) would be different?

14. Sound waves cause air to oscillate. What do light waves oscillate?

15. Order the following colors in terms of increasing wavelength: blue, red, yellow, green.

16. Do radio waves have longer or shorter wavelengths than visible light? What about X-rays?

17. In a physics experiment, a light ray is examined as it travels from air into glass. If the angle that the light ray makes with a line perpendicular to the glass surface is measured, will the refracted ray bend toward or away from that line?

18. A shirt is dyed so that it looks yellow. What colors of light does the dye absorb?

19. A cyan dye is made of a mixture of substances which absorb all light colors except blue and green. If you took a cyan piece of paper and placed it in a dark room and shined blue light on it, what would you see? What would you see if you shined yellow light on it?

20. From the inside to the outside, name the four regions of the sun.

21. How does the sun get its power? In which region of the sun does this process occur?

22. For both nuclear fusion and nuclear fission discussed in the text, what can we say about the mass of the starting materials compared to the mass of what's made in the end?

Problems 23 - 25 refer to the following H-R diagram:

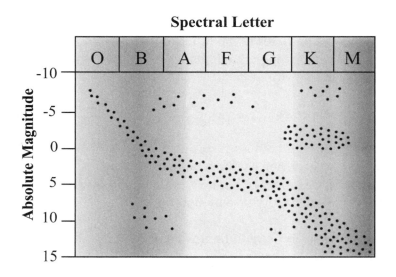

23. Classify a star with Magnitude -7, Spectral Letter F

24 A star much like our own sun has a Spectral Letter of G. Specify the minimum and maximum magnitude for this star.

25. As one moves from left to right on the H-R diagram, do the stars get warmer or cooler?

SOLUTIONS TO QUARTERLY TEST #1

1. (6 pts – one for each definition)

a. <u>Molecule</u> – Two or more atoms linked together to make a substance with unique properties

b. <u>Relative humidity</u> – The ratio of the mass of water vapor in the air at a given temperature to the maximum mass of water vapor the air could hold at that temperature, expressed as a percentage

c. <u>Barometer</u> - An instrument used to measure atmospheric pressure

d. <u>Jet streams</u> – Narrow bands of high-speed winds that circle the earth, blowing from west to east

e. <u>Solvent</u> – A liquid substance capable of dissolving other substances

f. <u>Cohesion</u> – The phenomenon that occurs when individual molecules are so strongly attracted to each other that they tend to stay together, even when exposed to tension

2. (1 pt – ½ for each) Volume is measured in <u>liters</u> in the metric system. In the English system, it is measured in <u>gallons, pints, or quarts</u> (the student need list only one of the English units).

3. (2 pts – one for the conversion relationship and one for the actual conversion) First, we convert the number to a fractional form:

$$\frac{12 \text{ m}}{1}$$

Next, since we want to convert from meters to centimeters, we need to remember that "centi" means "0.01." So one centimeter is the same thing as 0.01 meters. Thus:

$$1 \text{ cm} = 0.01 \text{ m}$$

That's our conversion relationship. Since we want to end up with cm in the end, we must multiply the measurement by a fraction that has meters on the bottom (to cancel the meter unit that is there) and cm on the top (so that cm is the unit we are left with). Remember, the numbers next to the units in the relationship above go with the units. Thus, since "m" goes on the bottom of the fraction, so does "0.01." Since "cm" goes on the top, so does "1."

$$\frac{12 \text{ \sout{m}}}{1} \times \frac{1 \text{ cm}}{0.01 \text{ \sout{m}}} = 1,200 \text{ cm}$$

Thus, 12 m is the same as <u>1,200 cm</u>.

4. (2 pts – one for the conversion relationship and one for the actual conversion) First, we convert the number to a fractional form:

$$\frac{75 \text{ g}}{1}$$

Next, since we want to convert from grams to kg, we need to remember that "kilo" means "1,000." So one kilogram is the same thing as 1,000 grams. Thus:

$$1 \text{ kg} = 1,000 \text{ g}$$

That's our conversion relationship. Since we want to end up with kg in the end, we must multiply the measurement by a fraction that has grams on the bottom (to cancel the g unit that is there) and kg on the top (so that kg is the unit we are left with):

$$\frac{75 \text{ g}}{1} \times \frac{1 \text{ kg}}{1,000 \text{ g}} = 0.075 \text{ kg}$$

The object's mass is <u>0.075 kg</u>.

5. (2 pts – one for a fraction that has "1" and "3.78" in it and one for the answer)

$$\frac{0.5 \text{ L}}{1} \times \frac{1 \text{ gal}}{3.78 \text{ L}} = 0.13 \text{ gal}$$

There are <u>0.13 gal</u> in half a liter. Your student may have rounded in a different place. That's fine.

6. (1 pt) <u>The difference is the concentration of the active ingredient</u>. The concentration of any chemical affects how it works. In this case, the more concentrated the active ingredient, the better the cleaner.

7. (2 pts – one for what and one for why) <u>The greenhouse effect is the process by which certain gases trap energy that is radiated from the earth. Without the greenhouse effect, the earth would be far too cold to support life.</u>

8. (1 pt) <u>The water will not evaporate</u>. Since the relative humidity is 100%, the air cannot hold any more water vapor. As a result, no water will evaporate from the glass.

9. (1 pt) <u>Sweat cools you off because when it evaporates, it takes energy from your skin</u>. When energy leaves your skin, it gets cooler.

10. (1 pt) <u>Nitrogen makes up the majority of the air we inhale</u>. See Figure 2.5.

11. (1 pt) <u>We exhale more oxygen</u>. See Figure 2.5.

12. (1 pt) <u>The air is much cleaner today than 30 years ago</u>. See Figure 2.9.

13. (1 pt) <u>The regulations that limit ozone production are designed to reduce ground-level ozone, because ozone is poisonous for us to breathe.</u> We want to increase the ozone levels in the ozone layer, where no one is breathing, but we want to decrease ozone anywhere that people breathe.

14. (1 pt) <u>An atmospheric pressure of 31.1 inches of mercury would be reported</u>. Since 1.0 atm corresponds to the average, sea-level value of atmospheric pressure, 1.04 atms means that the atmospheric pressure is higher than average.

15. (1 pt) <u>The first came from the homosphere</u>. In the homosphere, the mixture of gases in the air is the same throughout. It is the mixture we learned in Module #2. The heterosphere has many different compositions, depending on altitude.

16. (1 pt) <u>Troposphere, stratosphere, mesosphere</u>

17. (2 pts – ½ each for troposphere and mesosphere and one for stratosphere) <u>In the troposphere and mesosphere, the temperature decreases with increasing altitude, while just the opposite occurs in the stratosphere.</u>

18. (1 pt) The majority of earth's weather occurs in the <u>troposphere</u>.

19. (1 pt) The molecules move more slowly in the sample that has the lower temperature, which is the <u>second sample</u>.

20. (1 pt) <u>The "ozone hole" is caused by certain ozone-destroying agents which work only in the presence of the Polar Vortex. Since the Polar Vortex happens only in Antarctica and only during one part of the year, the ozone hole forms only during one part of the year and mostly over Antarctica.</u>

21. (1 pt) <u>CH_4</u>.

22. (1 pt) There is a subscript of 6 after the C, indicating six atoms there. The next letter is capital, so it must represent another atom. There is a subscript of 12 after it, indicating 12 of those. The next capital letter has a subscript of 6, indicating 6 more atoms. Thus, there are a total of <u>24</u> atoms in the molecule.

23. (1 pt) <u>The molecule will be nonpolar</u>. If the atoms all pull on electrons with the same strength, none will be able to get more than its fair share.

24. (1 pt) Carbon disulfide must be <u>nonpolar</u>, otherwise it would have dissolved in water.

25. (1 pt) <u>Hydrogen bonding</u> is responsible. Hydrogen bonding brings the molecules close together and makes them want to stay close together, but that can only happen in the liquid state.

Total possible points: 35

SOLUTIONS TO QUARTERLY TEST #2

1. (7 pts – one for each definition)

a. Transpiration – Evaporation of water from plants

b. Adiabatic cooling – The cooling of a gas that happens when the gas expands with no way of getting more energy

c. Sedimentary rock – Rock formed when chemical reactions cement sediments together, hardening them

d. Aphelion – The point at which the earth is farthest from the sun

e. Perihelion – The point at which the earth is closest to the sun

f. Coriolis effect – The way in which the rotation of the earth bends the path of winds, sea currents, and objects that fly through different latitudes

g. Insulator – A substance that does not conduct electricity very well

2. (1 pt) Since the vast majority of water on the earth is in the oceans, it is most likely to be saltwater.

3. (1 pt) If evaporation is the only way of getting rid of water, the salts that the lake receives will become concentrated, making it a saltwater lake.

4. (1 pt) The oceans are not salty enough for the earth to be billions of years old. Since salt accumulates in the oceans, the older the earth is, the saltier the oceans will be. Calculations indicate that even assuming the oceans had no salt to begin with, it would take only 62 million years (*not billions of years!*) to make the oceans as salty as they are now.

5. (1 pt) Glaciers start on mountains as the result of snow that never completely melts in the summer. If enough snow piles up, the weight causes it to slide down the mountain as a glacier.

6. (1 pt) The air will cool as it expands. That's what adiabatic cooling is all about.

7. (5 pts – one for each answer) The earth is divided into the atmosphere, hydrosphere, crust, mantle, and core.

8. (1 pt) Scientists observe seismic waves which are usually generated by earthquakes. The behavior of these waves tells us a lot about the makeup and properties of the mantle and core.

9. (1 pt) The magnetic field is caused by a large amount of electrical flow in the core.

10. (2 pts – ½ for each description and one for which is more valid) The dynamo theory says that the motion of the core is due to temperature differences in the core and the rotation of the earth. This motion causes the motion of electrical charges in the core, which creates electrical current. The rapid-decay theory states that the electrical current in the core started as a consequence of how the earth

formed and is reducing over time. The rapid-decay theory is more scientifically valid, since it is most consistent with the data.

11. (1 pt) Without the magnetic field, cosmic rays from the sun would hit the earth. These rays would kill all life on the planet.

12. (1 pt) Pangaea is a hypothetical supercontinent that might have existed in earth's past. At one time, all of the continents might have fit together to form this supercontinent.

13. (2 pts – ½ for each) The four types of mountains are: volcanic mountains, domed mountains, fault-block mountains, and folded mountains.

14. (3 pts – one for each) The three main factors are thermal energy, uneven distribution of energy, and water vapor in the atmosphere.

15. (1 pt) Incoming solar radiation

16. (1 pt) The Northern Hemisphere is pointed toward the sun at aphelion, so that's when it's summer in that hemisphere.

17. (1 pt) Temperature differences cause winds.

18. (1 pt) This kind of weather is indicative of a warm front. It is not a stationary front because the rain would have lasted several days.

19. (1 pt) This kind of cloud pattern and resulting rain is indicative of a cold front. Thus, the temperature should decrease after the rain.

20. (1 pt) The thunderstorm is probably made up of several cells. The mature stage of a typical thunderstorm cell lasts no longer than 30 minutes.

21. (1 pt) The return stroke is responsible for the majority of light and sound in a lightning bolt.

22. (1 pt) Thunder is the result of superheated air traveling out from the lightning bolt in waves. When those waves hit our eardrum, we interpret them as sound. Since the waves are violent, the sound is loud.

23. (1 pt) Sheet lightning is cloud-to-cloud lightning, while lightning bolts are cloud-to-ground lightning. The lightning bolts, therefore, hit the ground while sheet lightning never does.

24. (1 pt) A cumulonimbus cloud must be present to form a tornado. The vortex will not form without the strong updraft of a thunderstorm cell that forms a cumulonimbus cloud.

25. (1 pt) The Coriolis effect causes hurricanes in different hemispheres to rotate differently.

Total possible points: 39

SOLUTIONS TO QUARTERLY TEST #3

1. (8 pts – one for each definition)

a. Vector quantity – A physical measurement that contains directional information

b. Scalar quantity – A physical measurement that does not contain directional information

c. Acceleration – The time rate of change of an object's velocity

d. Kinetic friction – Friction that opposes motion once the motion has already started

e. Static friction – Friction that opposes the initiation of motion

f. Photon – A small "package" of light that acts like a particle

g. Electrical current – The amount of charge that travels past a fixed point in an electric circuit each second

h. Resistance – The ability of a material to impede the flow of charge

2. (2 pts – one for the conversion and one for the answer) This problem gives us distance and time and asks for speed. Thus, we need to use Equation (9.1). The problem wants the answer in miles per hour, however. We are given the time in minutes. Thus, we must make a conversion first:

$$\frac{45 \text{ minutes}}{1} \times \frac{1 \text{ hour}}{60 \text{ minutes}} = 0.75 \text{ hours}$$

Now we can use our speed equation:

$$\text{speed} = \frac{20 \text{ miles}}{0.75 \text{ hours}} = 26.7 \ \underline{\frac{\text{miles}}{\text{hour}}}$$

3. Vector quantities contain information about direction, scalar quantities do not. The units tell you what the measurement is of:

a. (1 pt – ½ for each answer) The units (distance divided by time^2) means this is acceleration. Since there is a direction, this is a vector quantity.

b. (1 pt – ½ for each answer) The units (distance divided by time) means this is either speed or velocity. Since there is no direction, it is a scalar quantity and is a measurement of speed.

c. (1 pt – ½ for each answer) The units (distance divided by time) means this is either speed or velocity. Since there is no direction, it is a scalar quantity and is a measurement of speed. Even though the information about slowing is in the measurement, that tells us nothing about direction.

d. (1 pt – ½ for each answer) The units (distance divided by time) mean this is either speed or velocity. Since there is a direction, it is a <u>vector quantity</u> and is a measurement of <u>velocity</u>.

4. (1 pt) The initial velocity is 0, and the final velocity is 15 meters per second east. The time is 2.1 seconds. This is a straightforward application of Equation (9.2).

$$\text{acceleration} = \frac{\text{final velocity} - \text{initial velocity}}{\text{time}}$$

$$\text{acceleration} = \frac{15\ \dfrac{\text{meters}}{\text{second}} - 0\ \dfrac{\text{meters}}{\text{second}}}{2.1\ \text{seconds}} = \frac{15\ \dfrac{\text{meters}}{\text{second}}}{2.1\ \text{seconds}} = 7.14\ \frac{\text{meters}}{\text{second}^2}$$

Since the acceleration is positive, it is in the same direction as the velocity. Thus, the acceleration is <u>7.14 m/sec^2 east</u>.

5. (1 pt) The rock is in free fall, so we can use Equation (9.3). Since the problem wants the answer in meters, we need to use 9.8 meters per second^2 as the acceleration.

$$\text{distance} = \frac{1}{2} \times (\text{acceleration}) \times (\text{time})^2$$

$$\text{distance} = \frac{1}{2} \times (9.8\ \frac{\text{meters}}{\text{second}^2}) \times (3.8\ \text{seconds})^2 = \frac{1}{2} \times (9.8\ \frac{\text{meters}}{\text{second}^2}) \times (3.8\ \text{seconds}) \times (3.8\ \text{seconds})$$

$$\text{distance} = \frac{1}{2} \times (9.8\ \frac{\text{meters}}{\cancel{\text{second}^2}}) \times (14.44\ \cancel{\text{second}^2}) = \underline{70.756\ \text{meters}}$$

6. (1 pt – ½ for the value and ½ for the direction) As the picture shows, the car is behind the truck, but they are both traveling in the same direction. Thus, we get their relative velocity by subtracting their individual velocities:

relative velocity = 42 miles per hour - 37 miles per hour = 5 miles per hour

Since the car is traveling slower than the truck, the truck is pulling away from the car. Thus, the relative velocity is <u>5 miles per hour away from each other</u>.

7. (1 pt) Since the velocity is not changing, <u>the acceleration is zero</u>. The time was just put in there to fool you. Remember, acceleration is the change in velocity. With no change in velocity, there is no acceleration.

8. (1 pt) <u>The beanbag will not fall next to the tree. Instead, it will fall north of the tree</u>. This is once again an application of Newton's First Law. While it is in the boy's hand, the beanbag has a velocity going north. When the boy drops the beanbag, it will still have a velocity going north. Thus, as it falls, it will travel north. When it lands, then, it will be north of the tree. In fact, ignoring air

resistance, when it hits the ground, it will be right next to wherever the boy is at that instant, because it will be traveling north with the boy's velocity.

9. (1 pt) Since the object is moving with a constant velocity, the acceleration is zero. Since the total force exerted on an object is equal to the object's mass times its acceleration (Newton's Second Law), the total force on the object is zero as well. This means that the child exerts just enough force to counteract kinetic friction, but no more. We must be talking about kinetic friction because the toy is already moving. Thus, the child exerts a force of 12 Newtons to the east.

10. (1 pt) Ignoring friction, the only force involved is the force that the father exerts. Thus, the acceleration is due entirely to that force, and Equation (10.1) will give us the force's strength:

$$\text{total force} = (\text{mass}) \cdot (\text{acceleration})$$

$$\text{total force} = (22 \text{ kg}) \cdot \left(2.5 \frac{\text{m}}{\text{sec}^2} \right) = 55 \text{ Newtons}$$

Since the father's force is the only one in play, the father is pushing with a force of 55 Newtons north.

11. (1 pt) To get the box moving, the worker need only overcome the static frictional force. Thus the worker must exert slightly more than 500 Newtons.

12. (1 pt) The wall exerts a force of 20 Newtons west, because it is equal and opposite of the man's force.

13. (1 pt) The static frictional force is always greater than the kinetic frictional force. Thus, the static frictional force is 225 Newtons, and the kinetic frictional force is 100 Newtons.

14. (3 pts – ½ for each force and one for the two that are difference facets) The four fundamental forces are the gravitational force, the weak force, the strong force, and the electromagnetic force. The electromagnetic force and the weak force are really different facets of the same force. Thus, some say there are only 3 fundamental forces in creation.

15. (2 pts – one for each answer) The weakest force is the gravitational force. The strongest one is the strong force.

16. (1 pt) The only difference is that the distance between the objects was multiplied by 2. The gravitational force decreases when the distance between the objects increases. It is decreased according to the square of that increase. Thus, the force is divided by 2^2, which is 4. The new gravitational force, then, is 4 times smaller than the old one.

17. (2 pts – one for each arrow) Traveling from "A" to "B" tells us that it is traveling clockwise. Its velocity is straight in the clockwise direction. Since it is traveling at a constant speed, the only force is centripetal, which always points to the center of the circle

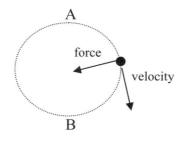

18. (2 pts – ¼ for each planet)

    Inner Planets - <u>Mercury, Venus, earth, and Mars</u>

    Outer Planets - <u>Jupiter, Saturn, Uranus, Neptune</u>

19. (2 pts – 1/3 for each part, and one for the part always present) A comet is made of <u>a nucleus, a coma, and a tail. The nucleus is always present.</u>

20. (2 pts – one for each answer) A low voltage means that the <u>energy of the individual electrons is low</u>. A large current means <u>there are many electrons flowing through the circuit.</u>

21. (1 pt) Conventional current flows from the positive side of a battery to the negative side.

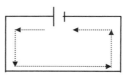

22. (1 pt) The wider, or thicker, the wire, the more the electrons can spread out. This means there is less likelihood of electrons colliding with atoms in the wire. Thus, the thicker wire will have less resistance. Less resistance means less heat. Thus, the <u>thin wire</u> will heat up the fastest.

23. (1 pt) <u>The bulbs are wired in parallel</u>, because a burnt-out bulb acts like an open switch. If the bulbs were wired in series, one open switch would stop any current from flowing. Since that does not happen, the bulbs must be wired in parallel, where one open switch will not affect the other parallel parts of the circuit.

24. (1 pt) <u>The flow of charged particles</u> causes a magnetic field.

25. (1 pt) <u>No</u>. As far as we know, magnets must always have both a north and south pole.

Total possible points: 43

SOLUTIONS TO QUARTERLY TEST #4

1. (11 pts – one for each definition)

a. <u>Mass number</u> – The sum of the numbers of neutrons and protons in the nucleus of an atom

b. <u>Isotopes</u> – Atoms with the same number of protons but different numbers of neutrons

c. <u>Atomic number</u> – The number of protons in an atom

d. <u>Transverse wave</u> – A wave with a direction of propagation that is perpendicular to its direction of oscillation

e. <u>Longitudinal wave</u> – A wave with a direction of propagation that is parallel to its direction of oscillation

f. <u>Supersonic speed</u> – Any speed that is faster than the speed of sound in the substance of interest

g. <u>Electromagnetic wave</u> – A transverse wave composed of an oscillating electric field and a magnetic field that oscillates perpendicular to the electric field

h. <u>The Law of Reflection</u> – The angle of reflection equals the angle of incidence.

i. <u>Nuclear fusion</u> – The process by which two or more small nuclei fuse to make a bigger nucleus

j. <u>Nuclear fission</u> - The process by which a large nucleus is split into two smaller nuclei

k. <u>Light year</u> – The distance light could travel along a straight line in one year

2. (2 pts – 1/3 for each particle and one for the proper order) The three constituents of the atom are the proton, neutron and electron. The electrons are significantly smaller than the other two, and the neutron is just slightly heavier than the proton. Thus, the order is <u>electron, proton, neutron</u>.

3. (1 pt – ½ for each answer) <u>The nuclear force</u> holds the protons and neutrons in the nucleus. The student could have called it the strong force as well. <u>This force is caused by the exchange of pions between protons and/or neutrons</u>.

4. (1 pt – 1/3 for each answer) Since the chemical symbol is C, we can use the chart to learn that the atom has <u>6 protons</u>. This tells us there are also <u>6 electrons</u>. The mass number is the sum of protons and neutrons in the nucleus. Thus, there are <u>8 neutrons</u>.

5. (1 pt) ^{131}I has 53 protons according to the chart. This means there must be 78 neutrons. In beta decay, a neutron turns into a proton. This will result in an atom with 54 protons and 77 neutrons, or ^{131}Xe.

6. (1 pt) In 1,600 years, the 10 grams will be cut in half to 5 grams. In the next 1,600 years, that 5 grams will be cut in have to 2.5 grams. In the next 1,600 years, that 2.5 grams will be cut in half to 1.25 grams. That's a total of 4,800 years, so the answer is <u>1.25 grams</u>.

7. (1 pt) <u>Alpha particles pass through the least amount of matter before stopping, beta particles are next, and gamma rays pass through the most matter before stopping</u>.

8. (1 pt) The sound waves with the shortest wavelengths are <u>ultrasonic waves</u>. Remember, ultrasonic waves have the highest frequencies, which means the smallest wavelengths.

9. (1 pt – ½ for the fact that the alarm won't be head and one for why) <u>The astronaut will not be able to hear the alarm because there is no air on the surface of the moon. Thus, there is nothing that the sound waves can oscillate</u>.

10. (1 pt) Sound waves are <u>longitudinal waves</u>.

11. (2 pts – one for the speed and one for the answer) To determine the distance, we will use the time difference between the lightning flash and the sound. We will assume that the light from the lightning reaches our eyes essentially at the same time as the lightning was formed. Thus, the time it takes for the sound to travel to you will determine the distance. First, then, we need to know the speed of sound:

$$v = (331.5 + 0.6 \cdot T)\,\frac{m}{sec}$$

$$v = (331.5 + 0.6 \cdot 13)\,\frac{m}{sec}$$

$$v = (331.5 + 7.8)\,\frac{m}{sec} = 339.3\,\frac{m}{sec}$$

Now we can use Equation (14.3):

$$distance\ traveled = (speed) \times (time\ traveled)$$

$$distance\ traveled = (339.3\frac{m}{sec}) \times (1.5\,sec) = \underline{508.95\ m}$$

12. (1 pt) <u>When a jet travels at Mach 1 or higher, it emits a shock wave of air that causes a very loud boom. This boom can damage people's ears and buildings</u>.

13. (2 pts – ¼ for each answer) <u>The wavelength, frequency, and speed will all be the same</u>. After all, the pitch is determined by the frequency, which, in turn, determines the wavelength. The speed depends only on the temperature. <u>The amplitudes of the waves will be different</u>, however, because amplitude determines loudness.

14. (1 pt) <u>Light waves oscillate a magnetic field and an electrical field</u>. Each one oscillates perpendicular to the other, as well as perpendicular to the direction of travel.

15. (2 pts – ½ for each) The acronym ROY G. BIV allows us to remember the relative size of the colors' wavelengths. Red is longest and violet is shortest. Thus, in terms of increasing wavelength, the colors are: <u>blue, green, yellow, and red</u>.

16. (1 pt – ½ for each) <u>Radio waves have wavelengths longer than visible light, while X-rays have shorter wavelengths</u>.

17. (1 pt) When light travels from a substance in which it moves quickly to a substance in which it moves slowly, the light bends towards the perpendicular. Since light moves faster in air than in glass, <u>the light will bend towards the perpendicular line</u>.

18. (1 pt) In order to look yellow, it must absorb all colors except yellow. <u>Thus, it absorbs red, orange, green, blue, indigo, and violet light</u>.

19. (1 pt – ½ for each) Since cyan absorbs all colors except blue and green, it will absorb any yellow light shone on it. As a result, no light will make it to your eyes. <u>In yellow light, then, the paper would look black</u>. When you shined blue light on it, the blue light would be reflected. There would be no green to mix with it, though. <u>In blue light, the paper would look blue</u>.

20. (1 pt – ¼ for each) Starting on the inside, the sun is divided into <u>the core, the radiative zone, the convection zone, and the photosphere</u>.

21. (1 pt – ½ for each) <u>The sun gets its power from nuclear fusion that occurs in the core</u>.

22. (1 pt) In both processes, mass is converted into energy. Thus, <u>the mass of the starting materials is larger than the mass of the materials the process makes</u>.

23. (1 pt) To classify a star, you find where its magnitude and spectral letter put it on the H-R diagram.

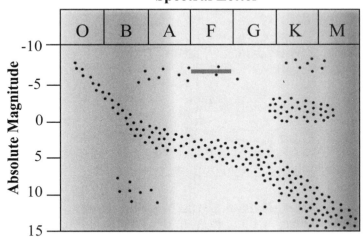

According to the magnitude and spectral letter, this star is a <u>supergiant</u>.

24. (1 pt) In order to be like our sun, it must be a main sequence star. Thus, it must fall in the main sequence group. In order to fall into the main sequence group with a spectral letter of G, it must have a magnitude <u>between 2 and 8</u>. The student's answers might be slightly different than mine, because he might read the H-R diagram slightly differently than I did.

25. (1 pt) The farther to the right on the H-R diagram, <u>the cooler the star</u>.

Total possible points: 39